Thomas F. Mütsch, Matthias B. Kowalski
Space Technology
De Gruyter Graduate

Also of interest

The Rocket into Planetary Space
Hermann Oberth, 2014
ISBN 978-3-486-75463-6, e-ISBN (PDF) 978-3-11-036756-0,
e-ISBN (EPUB) 978-3-486-99067-6

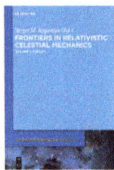

Frontiers in Relativistic Celestial Mechanics: Volume 1 Theory
Sergei M. Kopeikin (Ed.), 2014
ISBN 978-3-11-033747-1, e-ISBN (PDF) 978-3-11-033749-5,
e-ISBN (EPUB) 978-3-11-038938-8

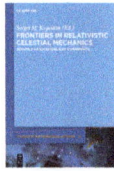

Frontiers in Relativistic Celestial Mechanics: Volume 2 Applications and Experiments
Sergei M. Kopeikin (Ed.), 2014
ISBN 978-3-11-034545-2, e-ISBN (PDF) 978-3-11-034566-7,
e-ISBN (EPUB) 978-3-11-037953-2

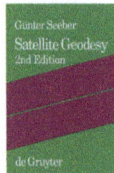

Satellite Geodesy: Foundations, Methods, and Applications
Günter Seeber, 2008
ISBN 978-3-11-017549-3, e-ISBN (PDF) 978-3-11-020008-9

Wind Energy Harvesting: Micro-to-Small Scale Turbines
Ravi Kishore, Stewart Priya, Colin Shashank, 2016
ISBN 978-1-61451-565-4, e-ISBN (PDF) 978-1-61451-417-6,
e-ISBN (EPUB) 978-1-61451-979-9

International Journal of Turbo & Jet-Engines
Benjamin Gal-Or (Editor-in-Chief)
ISSN 2191-0332

Thomas F. Mütsch, Matthias B. Kowalski

Space Technology

A Compendium for Space Engineering

DE GRUYTER

Authors
Dipl.-Ing. Thomas F. Mütsch
thomas.muetsch@freenet.de

Dr. Matthias B. Kowalski
space.book@gmx.net

ISBN 978-3-11-041321-2
e-ISBN (PDF) 978-3-11-041322-9
e-ISBN (EPUB) 978-3-11-042621-2

Library of Congress Cataloging-in-Publication Data
A CIP catalog record for this book has been applied for at the Library of Congress.

Bibliografische Information der Deutschen Nationalbibliothek
The Deutsche Nationalbibliothek lists this publication in the Deutsche Nationalbibliografie;
detailed bibliographic data are available on the Internet at http://dnb.dnb.de.

© 2016 Walter de Gruyter GmbH, Berlin/Boston
Cover image: 3DSculptor/iStock/thinkstock
Printing and binding: CPI books GmbH, Leck
♾ Printed on acid-free paper
Printed in Germany

www.degruyter.com

Preface

This textbook is a compendium for further education of students and jobholders in aerospace industry. For all other people who are interested in astronomy and astronautics this book should be also helpful for knowledge transfer. However, this book requires knowledge in higher mathematics and physics. Many of mathematical equations required are summarised in the appendix. On the derivation of analytical relations is generally waived. A realistic and practically oriented application is favoured. For private study and further learning a collection of questions is attached at the end of each chapter. An extensive collection of equations is given in the appendix.

The theoretical foundations of space flight had been developed in the last 20th century. The most important documents of this time are referenced in the text. Also important milestones in the practical implementation for the use in launcher systems and payloads to some current major projects on the further development of space flight are described. The book also covers the fundamentals of aerospace engineering. It explains the details of technical implementations organised in the border area of technical feasibility. Moreover, it explains the constraints of space flight and describes the key elements of rocket motors and power supply in more detail. The accessibility of celestial bodies is tabulated and documented in the outlook chapter, in which the largest vision of space flight, humans to Mars, is explained.

Space flight requires high precision, high development expenses, and therefore, high costs. Just a few countries are willing and able to afford the economic costs involved. Although it is 60 years since the first satellite was carried into space, annually less than 100 space missions are launched. This textbook is intended as a small contribution to the development and peaceful use of space flight.

Dipl.-Ing. Thomas F. Mütsch
Boxberg, Germany

Dr. rer. nat. Matthias B. Kowalski
Weimar, Germany

Cover illustration

The colour illustration on the cover shows NASA's Space Launch System (SLS). The preliminary design of the SLS was completed in 2013 and moved into production of the launch vehicle.

The SLS is an advanced launch vehicle for explorations beyond Earth's orbit into deep space. The world's most powerful rocket will launch astronauts in the new Orion spacecraft on missions to asteroids and eventually to planet Mars. Offering the highest-ever payload mass, this launcher also opens new possibilities for robotic scientific missions to planets like Mars, Saturn, and Jupiter.

The lift capability of the SLS enables the launch of larger payloads than any other commercial launcher systems. The high performance reduces the time for the travel of robotic spacecrafts through the Solar System, and by reducing the costs and risks the SLS provides larger volumes due to larger payload fairings, to fly on science missions that are too large for other commercial launchers. There will be several versions of the rocket to fit NASA's needs for future deep space missions beginning with a 70 t lift capability to one of a 130 t lift capability.

Contents

1 Historical Background

The earliest thoughts of flights into space date already back to the beginning of the discovery of black powder by ancient Chinese pyrotechnicians. In the following centuries to early modern times most of the desires for takeoff and heading for the Moon were written down in works of fiction. However, the early physicists, mathematicians, and astronomers created the fundamentals for our conception of the world. Thus, they laid the cornerstones for successful space flight. From the fundamentals of rocket construction in the nineteenth century to modern aerospace transportation, important milestones of space flight were set by unique people.

Most important people of space flight

Konstantin Eduardovich Tsiolkovski (1857 – 1935)
1898: Fundamentals of rocket construction, rocket equation.
Hermann Oberth[1] (1894 – 1989)
Nineteen-twenties: Fundamentals of space technology.
Walter Hohmann[2] (1880 – 1945)
Nineteen-twenties: Calculation of satellite orbits.
Robert Hutchinson Goddard (1882 – 1945)
16 March 1926: Launch of the first liquid propellant rocket (petrol and oxygen).
Sergei Pavlovich Korolev (1906 – 1966)
1933: Development of jet engines for liquid propellant rockets.
Head of the Soviet space programme.
Wernher von Braun (1912 – 1977)
3 October 1942: First launcher/combat bomb of the world (A-4/V-2).
Until 1945: Director of the Army Rocket Centre at Peenemünde, Germany.
Nineteen-sixties: Director of the U.S. Lunar Landing Mission (Saturn V, Apollo).

Most important events of space flight

4 October 1957: First satellite in space (Sputnik, Soviet Union).
12 April 1961: First man in space (Jurij Gagarin, Major, 1934 – 1968).
20 July 1969: First man on the Moon (Armstrong/Aldrin, USA).
24 December 1979: First launch of Ariane rocket from Kourou, French Guiana.
12 April 1981: First reusable space shuttle (Space Shuttle, USA).
28 January 1986: First shuttle crash (Challenger, USA. Seven crew members died).
1 February 2003: Second shuttle crash (Columbia, USA. Seven crew members died).
21 July 2011: Last shuttle landing, 135 flights in total.

---

1 Oberth H. Wege zur Raumschiffahrt. Verlag von R. Oldenbourg, München and Berlin, 1929.
2 Hohmann W. Die Erreichbarkeit der Himmelskörper. Oldenbourg Verlag, München, 1925.

Figure 1.1: American astronaut Edwin E. Aldrin on the Moon (1969). Credit: NASA [1].

The first long-distance rocket Bumper 8 was successfully launched from Launch Complex 3 (LC-3) of Cape Canaveral Air Force Station (CCAFS) on the east coast of Florida, USA on 24 July 1950. The two-staged Bumper 8 employed a German V2 rocket as 1st stage and an American Without Any Control (WAC) Corporal sounding rocket as 2nd stage.

Almost 20 years later, the most prominent event in human space flight history was probably the first manned lunar landing. Launched by a Saturn V rocket from Kennedy Space Center in Florida, the Apollo Lunar Module *Eagle* landed in the *Mare Tranquillitatis* on the Moon. Two Americans, *Neil A. Armstrong* and *Edwin E. Aldrin*, stayed a total of about 21.5 h on the lunar surface while *Michael Collins* piloted the Command/Service Module in the lunar orbit. All three spacemen returned to Earth and landed in the Pacific Ocean on 24 July 1969.

After more than forty years of further development, space flight reached the limits of its technical and area (astronomical) potential. Since the nineteen-eighties space flight became an established art of engineering embedded between many other special disciplines of mechanical engineering. Led by cost reductions and standardisation, more and more commercialisation and competition for market shares replace expensive new developments.

1.1 Consolidation of Space Flight

The consolidation of space flight is illustrated in Figure 1.2. Space flight is high energy consuming, cost-intensive, and risky. These factors prevent space flight from further rapid development, as has happened in general aviation since the first powered flight in 1903. To date, only three nations, Soviet Union/Russia, the United States, and China, realised manned space flight and six nations, the United States, Soviet Union/Russia, Europe, China, Japan, and India, developed competence to place geostationary satellites. Since the first launch of an A-4/V-2 rocket from Peenemünde in 1942, an increasing number of rockets have been launched over the last decades. The number of flights into orbit are shown in Figure 1.2.

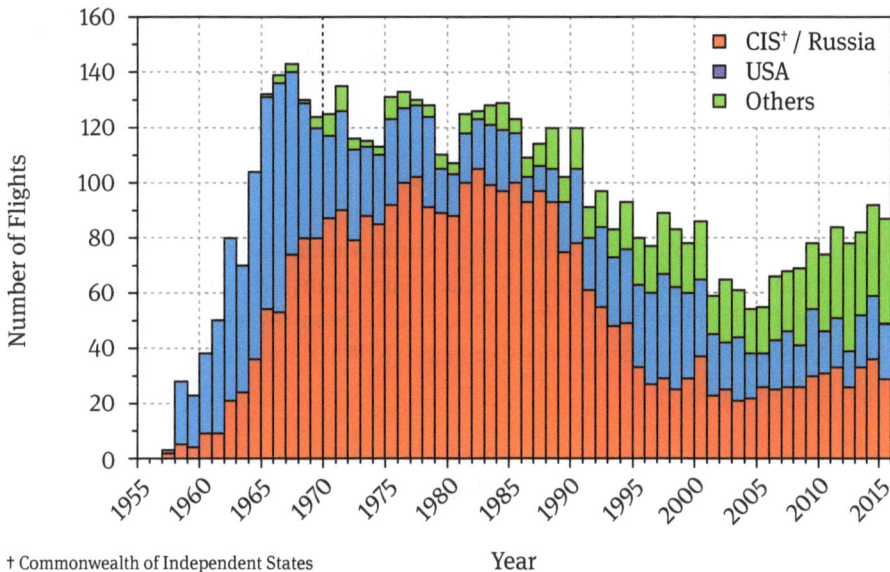

† Commonwealth of Independent States

Figure 1.2: Number of rocket launches including losses.

The end of the Cold War between the superpowers USA and Soviet Union led to a significant reduction of military payloads, as a result of which the number of launches roughly halved per year. Also the accident of the American Space Shuttle Challenger led to a severe setback for space flight.

Along with the increasing number of rocket launches the number of non-functional artificial objects also increased in the orbit. A total of approximately 150 million of man-made objects have gone into space. This space debris is produced by launchers with disposable boosters, payloads, discarded satellites, and mission-induced objects. Approximately 10 000 of these objects are fragments from numerous explosions and a few collisions in outer space.

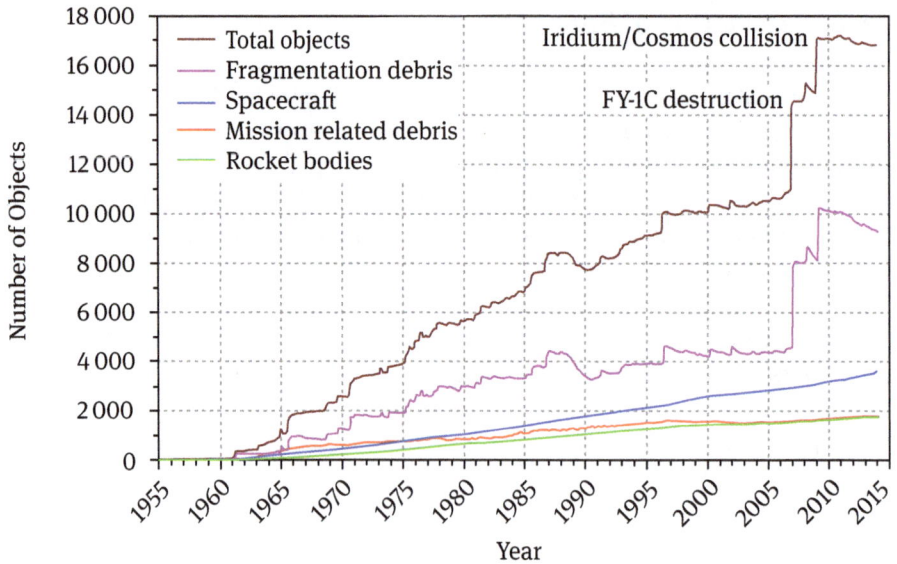

Figure 1.3: Number of objects in Earth orbit. Data: NASA [2].

Approximately 17 000 objects within Earth's orbit are currently tracked from ground and officially catalogued by the U.S. Space Surveillance Network (see Figure 1.3). Far beyond this, more than 41 000 objects have been registered in space by the North American Aerospace Defense Command (NORAD) from the beginning of space flight until today.

1.2 Questions for Further Studies

1. Give the names of the most important space pioneers.
2. Give the name and year of the first successfully launched long-range rocket.
3. What was the name of the first satellite launched into outer space?
4. Which country was the first in outer space?
5. Who was the first man in outer space?
6. What was the name of the spacecraft with the first man in outer space aboard?
7. Who was the first man on the Moon?
8. Give the names and year of the major accidents in the history of space flight?
9. What astronauts/cosmonauts were the most famous of all time?
10. Give the name of the American theoretical physicist who was involved in the investigation of the Space Shuttle Challenger disaster.

2 Basic Principles

For the determination of dimensions in astronautics, the knowledge of our astronomical environment is necessary. Limited by physical and technical feasibilities, our considerations should be restricted to travels within the borders of our solar system or within the close periphery of the solar system. At a distance of four light years, the next star Proxima Centauri is reached in a flight time of at least 20 000 years (!) with the today's technical means. Furthermore, we have to restrict our considerations to velocities very much lower than the speed of light. Therefore, all relativistic influences are negligible.

2.1 Solar System

Our planetary system consists of eight big planets and a great number of moons, asteroids (planetoids), plutoids, comets and other small bodies.

Table 2.1: Planets and moons of the Solar System.

Celestial body	Radius / m	Semi-major axis / m	Mass / m	Moons (not complete)
Sun	696 000E+3	–	1.9891E+30	–
Mercury	2 439E+3	57.91E+9	0.3302E+24	–
Venus	6 052E+3	108.21E+9	4.869E+24	–
Earth	6 371.2E+3	149.598E+9	5.974E+24	Earth's Moon
Mars	3 397E+3	227.94E+9	0.64219E+24	Phobos, Deimos
Jupiter	71 492E+3	778.4E+9	18.988E+26	Io, Europa, Ganymede, Callisto
Saturn	60 268E+3	1 427E+9	5.684E+26	Titan
Uranus	25 559E+3	2 870E+9	0.8683E+26	Oberon, Ariel
Neptune	24 766E+3	4 496E+9	1.0247E+26	Triton
Earth's Moon	1738E+3	384 403E+3	7.3432E+22	–

There are many mnemonic devices to remember the planets orbiting the Sun. Here is one among many others. The first letter of each word gives the first letter of the planets, in the order 'My Very Educated Mother Just Send Us Nine' without the dwarf planet Pluto at the end.

Table 2.2: Physical constants.

Constant	Symbol	Value	Unit
Speed of light	c	299 972.458E+3	$m \cdot s^{-1}$
Gravitational constant	G	6.672E−11	$m^3 (kg \cdot s^2)^{-1}$
Universal gas constant	R	8.31441	$J (K \cdot mol)^{-1}$
Solar constant	S	1372	$W \cdot m^{-2}$
Astronomical unit	AU	149.598E+9	m
Faraday constant	F	96 484.55	$A \cdot s \cdot mol^{-1}$
Permitivity	$\varepsilon 0$	$(36E+9 \cdot \pi)^{-1}$	$Farad \cdot m^{-1}$

2.2 Atmosphere of the Earth

The knowledge of the composition of the high atmosphere is necessary for space flight in the Low Earth Orbit (LEO). The diagrams show the course of the molecular weight, the density and the static temperature as a function of altitude.

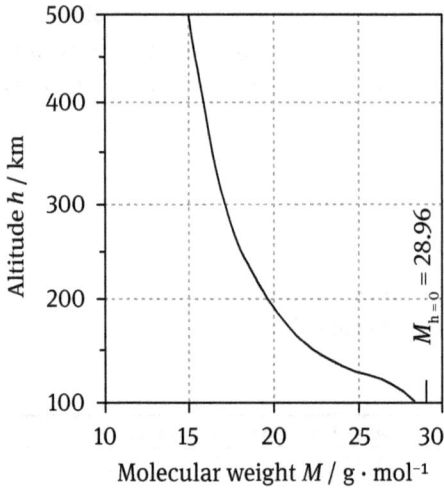

Figure 2.1: Average molecular weight M of Earth's high atmosphere. Data: COSPAR [3].

The range deviation in Figures 2.2 and 2.3 (curves with dashed lines) is due to the solar activity and the terrestrial magnetic field.

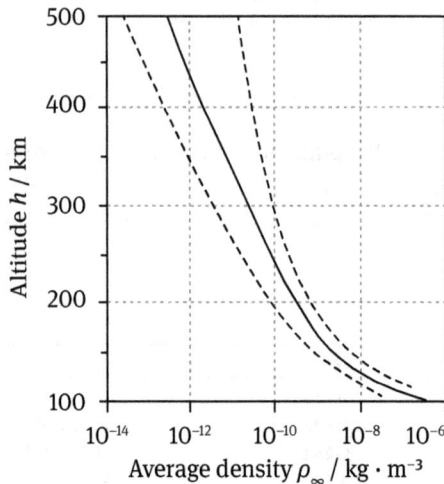

Figure 2.2: Average density ρ_∞ of Earth's high atmosphere. Data: COSPAR [3].

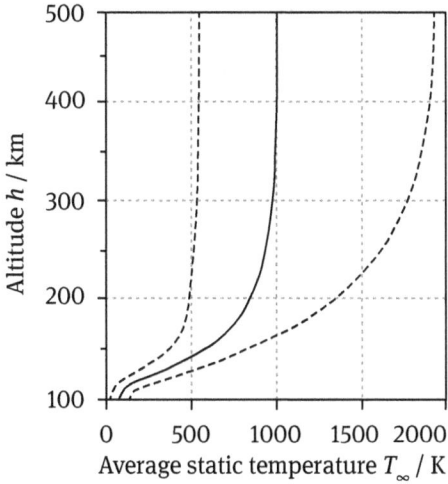

Figure 2.3: Average static temperature T_∞ of Earth's high atmosphere. Data: COSPAR [3].

2.3 Distances and velocities

Dealing with quantities in astronautics, the knowledge of distances and velocities in our macrocosm is of great importance. Some specific distances and linear and rotational velocities are exemplarily represented in the following tables below. The data are listed in scientific notation.

Table 2.3: Distances and velocities.

Altitude / Dia. / Distance / Height	Distance / m	Comment
Height of Mount Everest	8.85E+3	Above sea level
Altitude Airbus jetliner	11E+3	Roughly estimated
Altitude of Concorde jetliner	18E+3	Roughly estimated
Earth's atmosphere	100E+3	Roughly estimated
Altitude of the ISS	350E+3	Roughly estimated
Average diameter of the Earth	12742.4E+3	–
Maximum distance on Earth	20040E+3	Half of circumference of Earth
Geostationary orbit	35800E+3	Earth
Average distance Earth – Moon	384403E+3	–
Average diameter of the Sun	1392000E+3	–
Average distance Earth – Sun	149.6E+9	Astronomical unit AU
Average distance Earth – Pluto	5913.52E+9	–
One light year	9.461E+15	Per Julian year
Distance to the next star	40E+15	Proxima Centauri
Diameter of the Milky Way	1200000E+15	–
Maximum distance in the Universe	150000E+21	–

2.3.1 Linear Velocity

The linear velocity v is defined as the distance s travelled per unit of time t.

$$v = \frac{\Delta s}{\Delta t} \tag{2.1}$$

Linear velocities of different vehicles and objects compared with a human world record sprinter are shown in Table 2.4.

Table 2.4: Typical velocities.

Body	Linear velocity / $m \cdot s^{-1}$	Comment
100-m sprinter	10	World record
Vehicle	50	$180 \, km \cdot h^{-1}$
Airplane	300	Roughly estimated
Sonic velocity	330	Roughly estimated
Concorde jetliner	600	Roughly estimated
Bullet	800	Roughly estimated
Earth's Moon	1000	Around the Earth
Anti-tank shell	1800	Roughly estimated
Aggregate-4 missile	2800	Roughly estimated
Intercontinental ballistic missile	6000	ICBM
First cosmic velocity	7900	Earth
Second cosmic velocity	11200	Earth
Earth	30000	Around the Sun
Sun	300000	Around the galactic centre
Milky Way	600000	Within the local group
Speed of light	299792458	Probably the maximum possible speed

2.3.2 Angular velocity

The angular velocity w is defined as the rotation angle φ swept per unit of time t.

$$\omega = \frac{\Delta \varphi}{\Delta t} = \frac{2 \cdot \pi}{T} = 2 \cdot \pi \cdot v \tag{2.2}$$

Some rotational velocities of two different objects compared with the Earth are shown in Table 2.5.

Table 2.5: Rotational velocities.

Body	Rotational velocity / s^{-1}	Comment
Earth	7.3 E−5	$2\pi / 86164$
50 Hz generator	314	300 rpm
Flywheel	3140	30000 rpm

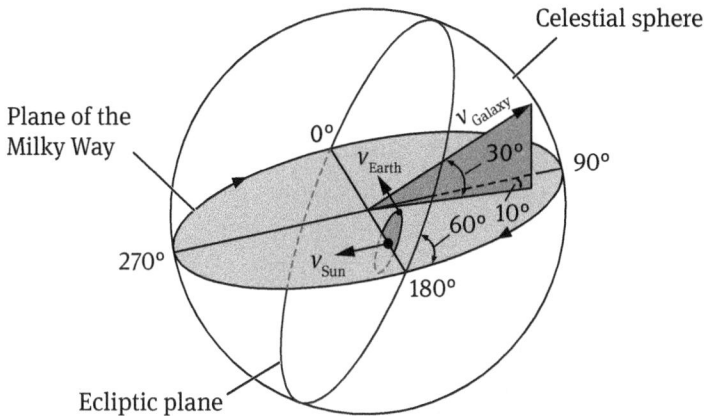

Figure 2.4: Position and velocities in the astronomical environment of the Earth.

Where

$v_{Galaxy} \approx 600 \, km \cdot s^{-1}$

$v_{Earth} \approx 30 \, km \cdot s^{-1}$

$v_{Sun} \approx 300 \, km \cdot s^{-1}$

2.4 Laws of Conservation of Energy and Momentum

The following elementary physical conservation equations are required for calculations in astronomy and astronautics.

Laws of Conservation of Energy

The total amount of energy in an isolated system is constant.
The energy in an isolated system can neither be created nor destroyed.
The energy in an isolated system can only be transformed from one state to the other.

The following forms of energy are significant in space technology.

Translational (kinetic) energy

$$E_{kin} = \frac{1}{2} m \cdot v^2 \tag{2.3}$$

Rotational energy

$$E_{rot} = \frac{1}{2} I \cdot \omega^2 \tag{2.4}$$

Potential energy

$$E_{pot} = -G\,\frac{M \cdot m}{r} \qquad (2.5)$$

Where

I is the mass moment of inertia
G is the gravitational constant
M is the mass of a natural body
m is the mass of an artificial body
r is the distance of the centres of mass

The mass moments of inertia I for different body shapes are listed in Table 2.6.

Law of Conservation of Linear Momentum

The linear momentum in an isolated system is constant.

$$\sum m \cdot v = \text{const.} \qquad (2.6)$$

Law of Conservation of Angular Momentum

The angular momentum in an isolated system is constant.

$$\sum J \cdot \omega = \text{const.} \qquad (2.7)$$

This gives the Law of the Lever

$$r_{min} \cdot v_{max} = r_{max} \cdot v_{min} \qquad (2.8)$$

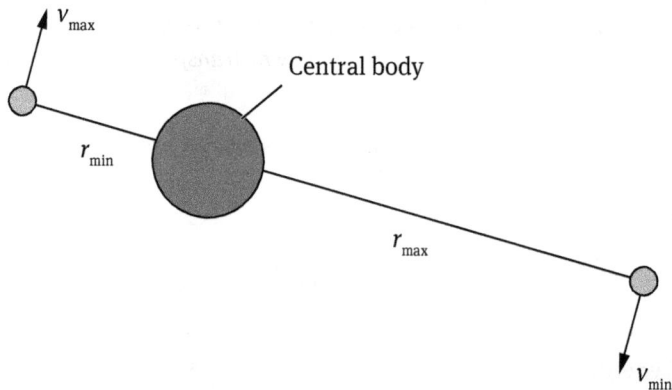

Figure 2.5: Law of the Lever.

Table 2.6: Mass moments of inertia of different homogeneous mass bodies

Solid cylinder

$$I_x = \frac{m(R^2)}{2}$$
$$I_y = \frac{m(3R^2 + h^2)}{12}$$
$$I_z = \frac{m(3R^2 + h^2)}{12}$$

with $m = \pi\rho R^2 h$

Hollow cylinder

$$I_x = \frac{m(R^2 + r^2)}{2}$$
$$I_y = \frac{m\left(R^2 + r^2 + \frac{h^2}{3}\right)}{4}$$
$$I_z = \frac{m\left(R^2 + r^2 + \frac{h^2}{3}\right)}{4}$$

with $m = \pi\rho\,(R - r)\,h$

Cuboid

$$I_x = \frac{m(b^2 + c^2)}{12}$$
$$I_y = \frac{m(a^2 + c^2)}{12}$$
$$I_z = \frac{m(a^2 + b^2)}{12}$$

with $m = \rho abc$

Thin rod

$$I_y = \frac{1}{12}\,ml^2$$
$$I_z = \frac{1}{12}\,ml^2$$

with $m = \rho A l$

Circular cone

$$I_x = \frac{3m\,R^2}{10}$$
$$I_y = \frac{3m\,(4R^2 + h^2)}{80}$$
$$I_z = \frac{3m\,(4R^2 + h^2)}{80}$$

with $m = \frac{\pi\rho R^2 h}{3}$

Pyramid

$$I_x = \frac{m(a^2 + b^2)}{20}$$
$$I_y = \frac{m\left(a^2 + \frac{3}{4}h^2\right)}{20}$$
$$I_z = \frac{m\left(b^2 + \frac{3}{4}h^2\right)}{20}$$

with $m = \frac{\rho\,ab\,h}{3}$

Sphere

$$I_x = \frac{2}{5}\,mR^2$$
$$I_y = \frac{2}{5}\,mR^2$$
$$I_y = \frac{2}{5}\,mR^2$$

with $m = \rho\pi\,\frac{4}{3}\,R^3$

Hollow sphere

$$I_x = \frac{2}{5}\,\frac{m(R^5 - r^5)}{(R^3 - r^3)}$$
$$I_y = \frac{2}{5}\,\frac{m(R^5 - r^5)}{(R^3 - r^3)}$$
$$I_y = \frac{2}{5}\,\frac{m(R^5 - r^5)}{(R^3 - r^3)}$$

with $m = \rho\pi\,\frac{4}{3}\,(R^3 - r^3)$

Hemisphere

$$I_x = \frac{83}{320}\,mR^2$$
$$I_y = \frac{83}{320}\,mR^2$$
$$I_z = \frac{2}{5}\,mR^2$$

with $m = \rho\pi\,\frac{2}{3}\,R^3$

Torus

$$I_x = \frac{m\,(4R^2 + 5r^2)}{8}$$
$$I_y = \frac{m\,(4R^2 + 5r^2)}{8}$$
$$I_z = \frac{m\,(4R^2 + 3r^2)}{4}$$

with $m = \rho\,2\pi^2 r^2 R$

Trucated cone

$$I_x = \frac{3}{10}\,\frac{m(R^5 - r^5)}{(R^3 - r^3)}\,h$$

with $m = \rho\,\frac{1}{3}\pi\,(R^2 + R\,r + r^2)\,h$

Rotation body

$$I_x = \frac{1}{2}\,\pi\rho \int_{x_1}^{x_2} f(x)^4\,dx$$

with $m = \pi\rho \int_{x_1}^{x_2} f(x)^2\,dx$

2.5 Theoretical Basics of Orbit Mechanics

In this textbook orbit mechanics is restricted to classical orbit mechanics or celestial mechanics. It is a special case of the more general relativistic orbit mechanics. Here, we consider for practical use only cases with velocities that are considerably lower than the speed of light. We do not make any significant mistake if we come to this restriction.

The basics of orbit mechanics were laid and explained for the first time by *Johannes Kepler* from the village of Weil der Stadt, Germany by the Laws of Planetary Motion, named after him. Kepler summarises the collected data of his mentor, *Tycho Brahe*, with three laws. These laws described the motion of planets in a solar system with a sun in its centre. Actually, Kepler's laws are still considered as an accurate description of the motion of any planet and any satellite revolving around the Sun or around the Earth.

Kepler's First Law (1609 A. D.)

The orbit of a planet is an ellipse with the Sun at one focus.

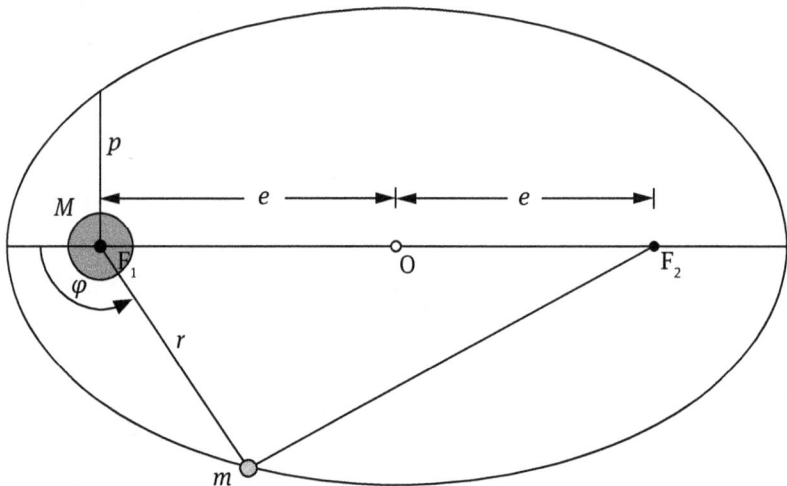

Figure 2.6: Kepler's First Law.

Kepler's First Law of Planetary Motion is also called the Law of Ellipses. By means of Newton's Law of Gravitation Kepler's First Law can be expressed in rather more general terms. During the motion of a celestial body within the gravitational field of a central mass, the orbit of the celestial body is running along a circle, an ellipse, a parabola or a hyperbola. These geometric shapes are conic sections and, in general, the orbits of any celestial bodies follow these conic sections (see Figure 2.7). Calculation in accordance with Equation 14.1.

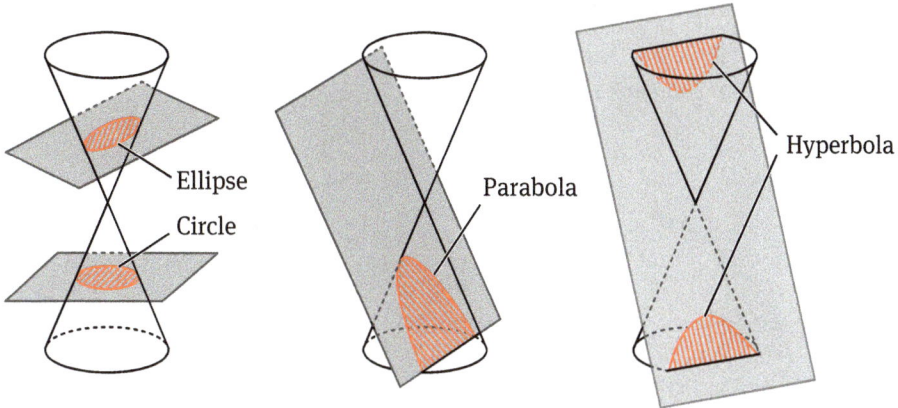

Figure 2.7: Schematic representation of different conic sections of circular cones.

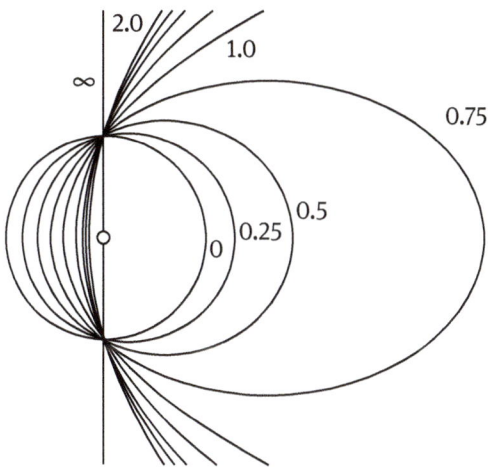

Figure 2.8: Geometric representation of conic sections with constant orbital parameter p.

In general, conic sections with an eccentricity of $e < 1$ (ellipse) and an eccentricity of $e > 1$ (hyperbola) occur in nature. For some applications the following cases are significant.

$e = 0$: Circular orbit
$e = 1$: Parabola
$e \rightarrow \infty$: Straight line

Any slight interferences, which are always present, prevent these exact mathematical equations from being accurate in nature and astronautics.

Kepler's Second Law (1609 A. D.)

The straight line or position vector joining the planet to the sun sweeps out equal areas in equal times.

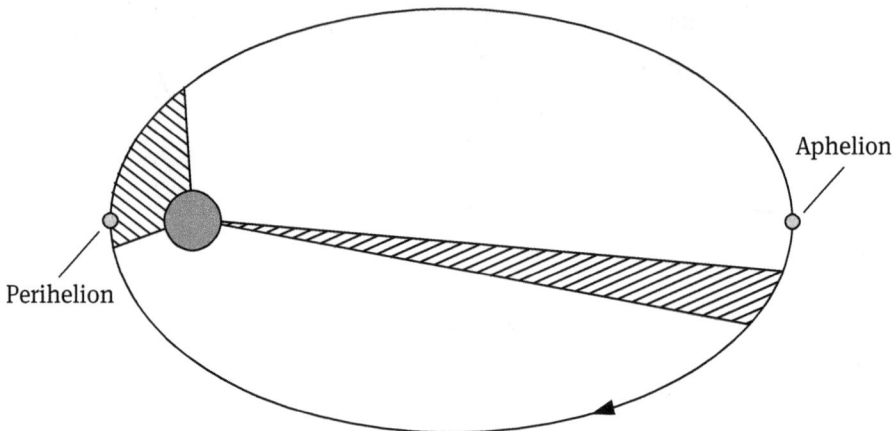

Figure 2.9: Kepler's Second Law.

Kepler's Second Law of Planetary Motion is also called the Law of Equal Areas because it explains that planets are orbiting the Sun in a path described as an ellipse. The velocity at which a planet moves around the Sun is constantly changing. The planet moves fastest when it is closest to the Sun and it moves slowest when it is furthest from the Sun. The point where the planet is nearest to the Sun is called *perihelion*. The opposite point is the *aphelion* where the planet is furthest from the Sun.

Kepler's Third Law (1618 A. D.)

The squares of the orbital periods of two planets are directly proportional to the cubes of the semi-major axes of their elliptical orbits.

$$\left(\frac{T_1}{T_2}\right)^2 = \left(\frac{a_1}{a_2}\right)^3 \tag{2.9}$$

Where

T_1 is the orbital period of the first planet
T_2 is the orbital period of the second planet
a_1 is the semi-major axis of first planet
a_2 is the semi-major axis of the second planet

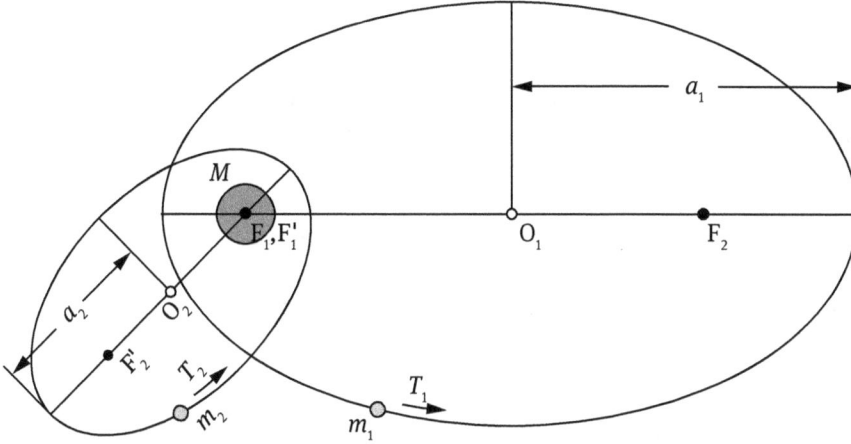

Figure 2.10: Schematic visualisation for understanding Kepler's Third Law.

Kepler's Third Law captures the relationship between the distance of a planet from the Sun, and its orbital period. The orbital period T of any celestial body on its path around the Sun can be calculated according to Equation 14.30. The prerequisite for this calculation is that the mass M of the Sun is much more bigger than the mass m of the planet ($M \gg m$).

As already mentioned above, *Isaac Newton* provided an explanation of Kepler's Laws of Planetary Motion by the Law of Universal Gravitation fifty years later (see Equation 2.10).

$$F = G \frac{M \cdot m}{r^2} \tag{2.10}$$

Regarding accelerations, with

$$F = a \cdot m \tag{2.11}$$

it follows that

$$a = \frac{F}{m} = \frac{G \cdot M}{r^2} \tag{2.12}$$

Where

F is the force of attraction
G is the gravitational constant
M is the mass of a natural body (e. g., the Sun)
m is the mass of an artificial body or planet
r is the distance of the centres of mass
a is the acceleration of gravity

2.5.1 Ballistic Trajectories

The ballistic trajectory is a special case of the more general elliptical orbit with the restriction that the peripoint of the trajectory is located within the radius of the central body or in the atmosphere above Earth's surface. Thus, no complete orbit is achieved. The concept of ballistics originates from military engineering. The terms internal ballistics (within the gun barrel or muzzle routed) and external ballistics are used. In classical orbit mechanics only external ballistics is the matter of interest after leaving the muzzle or burn-out of the missile. This results in the Equation 14.45 to Equation 14.52 neglecting the oblateness of the Earth and the air drag of the atmosphere.

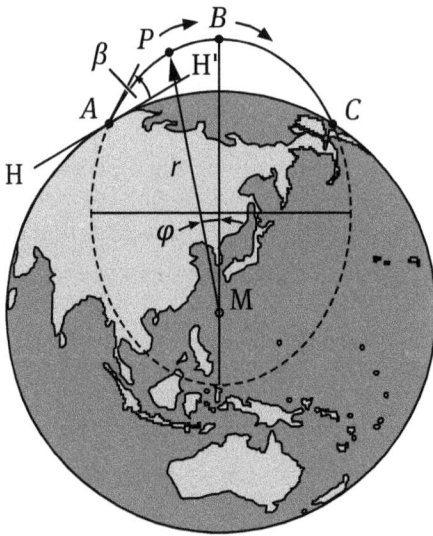

Figure 2.11: Schematic representation of a ballistic trajectory.

The ballistic trajectory of a missile as part of an imaginary ellipse with one focus in the centre of the Earth M is shown in Figure 2.11. The continues solid line above Earth's surface represents the real trajectory of a missile to the impact point in C. The dashed line is the imaginary trajectory if the mass of the Earth would be concentrated in M. The dashed line together with the continues solid line would then be the complete (non-ballistic) elliptic trajectory. Point P is any location of the missile on the ballistic trajectory with distance r to the centre of the Earth and the angle φ to the major-semi axis of the ellipse. At the starting point A the radius r is the idealised Earth radius R with an angle φ_0 (not shown), assuming the Earth is an ideal sphere.

The ballistic missile starts with a launching angle β, marked from the horizon $H-H'$ at point A, reaching its ceiling at point B, and continues the path to the point of impact C. For small launching angles the equations of the ballistic trajectories will be replaced by the equations of the oblique throw.

Range *s*

$$s = v^2 \sin(2\beta) \frac{R^2}{\mu}$$ (2.13)

Ceiling *h* (maximum altitude)

$$h = \frac{v^2 \sin^2 \beta}{2} \frac{R^2}{\mu}$$ (2.14)

Time of flight *t*

$$t = 2 v \sin \beta \frac{R^2}{\mu}$$ (2.15)

Where

v is the start velocity

β is the dropping inclination angle towards horizontal (launching angle)

R is the radius of the surface of the planet

μ is the standard gravitational parameter.

It is equal to $G \cdot M$

A computer generated geometric representation of a ballistic trajectory is shown in Figure 2.12.

1 Orbital plane with Earth
2 Ballistic trajectory
3 Position vector
4 Apo-circular orbit

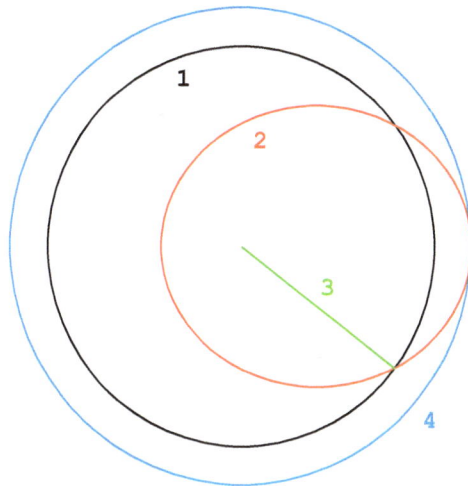

Figure 2.12: Geometric representation of a ballistic trajectory.

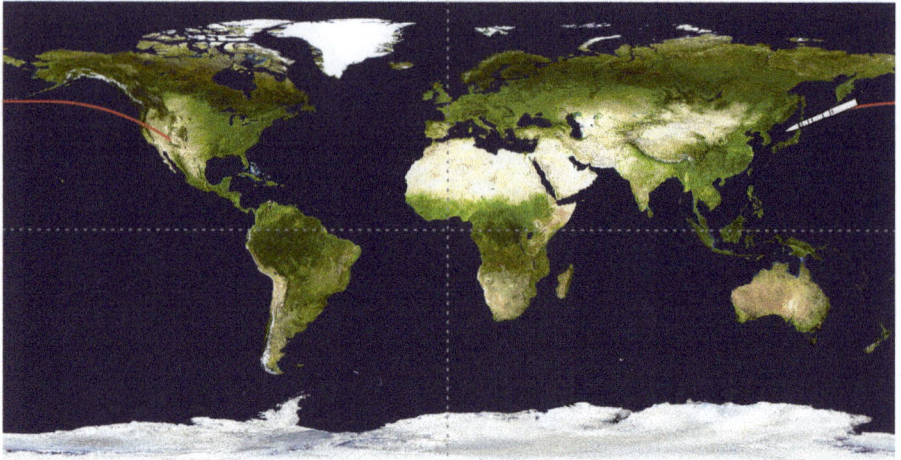

Figure 2.13: Ground track of a ballistic trajectory. Credit: NASA [4].

The ground track of a ballistic trajectory is shown in Figure 2.13. The computer calculated characteristic quantities of a ballistic trajectory from the Rocky Mountains in America to the Sea of Japan in Asia are listed below. The propulsion manoeuvres were neglected and the calculations were performed as free flight trajectory.

Computer generated data of the ground track for Earth

```
Flight time                    :  29m 38s
Apse angle (deg)               :  217.80
Latitude                       :  41.6.49
Lontitude                      :  139.57.24
Theor. impact velocity (m/s)   :  6886.59
```

2.5.2 Circular Orbits

The circular orbit is a special case of the more general elliptical orbit with two equal half-axes as boundary condition. Thus, the eccentricity of the ellipse vanishes ($e = 0$). In nature such mathematically accurate orbits do not exist. However, this approximate description is often used for representation. The shape of this orbit results from an orbit velocity that corresponds to the first cosmic velocity. The following equations are applied.

Geometric representation: Equations 14.1 to 14.10 (*see* section Geometry, Chapter 14.3, page 198)

Orbit determination: Equations 14.23 to 14.31 with $e = 0$ and $a = b = p = r$, respectively (*see* section Orbit determination, Chapter 14.3, pages 199 to 200)

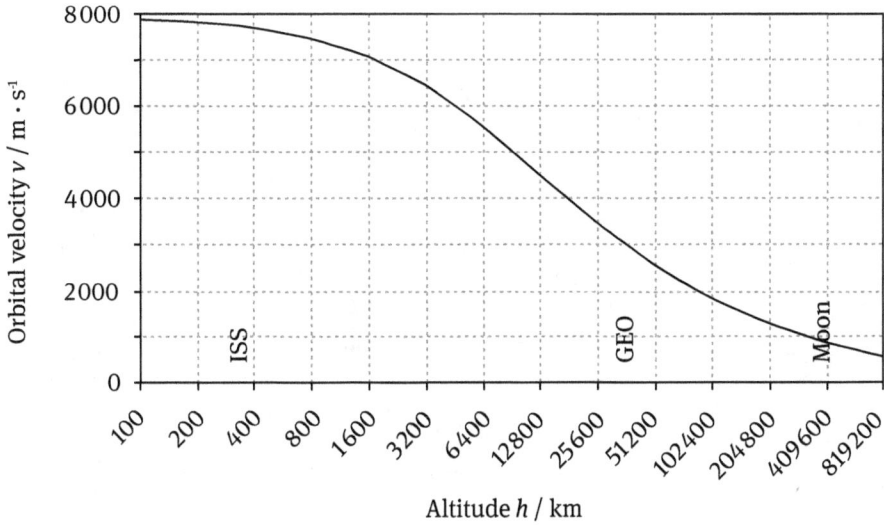

Figure 2.14: Velocity of circular orbits above Earth.

The constant velocity of a circular orbit and the orbital period of a natural and an artificial celestial body as a function of altitude above ground are shown in the Figures 2.14 and 2.15 exemplarily.

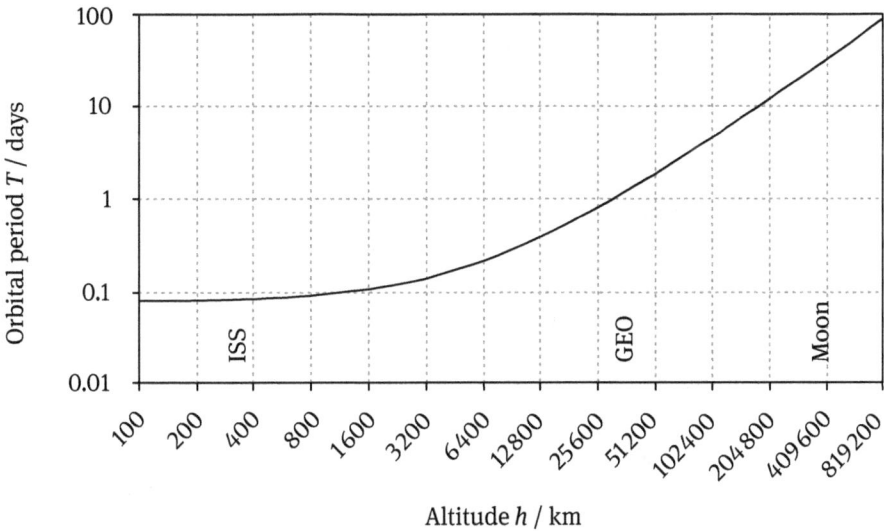

Figure 2.15: Orbital period on a circular orbit around Earth.

Low Earth Orbit

Figure 2.16: Russian Mir Space Station at an altitude of approximately 350 km. Credit: NASA [5].

The low Earth orbit (LEO) at an altitude of 250 – 1000 km is the simplest accessible orbit. It is the only possible orbit which is approached by manned missions (Space Shuttle or Soyuz), with the exception of flights to the Moon between 1968 and 1972.

The permanently manned International Space Station moved at an altitude of approximately 350 km above Earth with an inclination of 55° relative to the equator. Also the Russian Mir Space Station moved on a low Earth orbit around Earth. The photograph of the Russian Mir Space Station in Figure 2.16 was taken by an electronic still camera (ESC) from Space Shuttle Atlantis; STS-79; after undocking and separation from the space station during flight day 9.

The Russian Mir Space Station was designed for scientific research. Owned by the Soviet Union and later by Russia it was assembled in orbit from 1986 to 1996. The Mir was a microgravity research laboratory, in which the crews performed experiments in biology and physics to show influences of the reduced gravity on the experiments, and in astronomy and meteorology to take advantage of the absence of influences of Earth's atmosphere when taking observations and measurements.

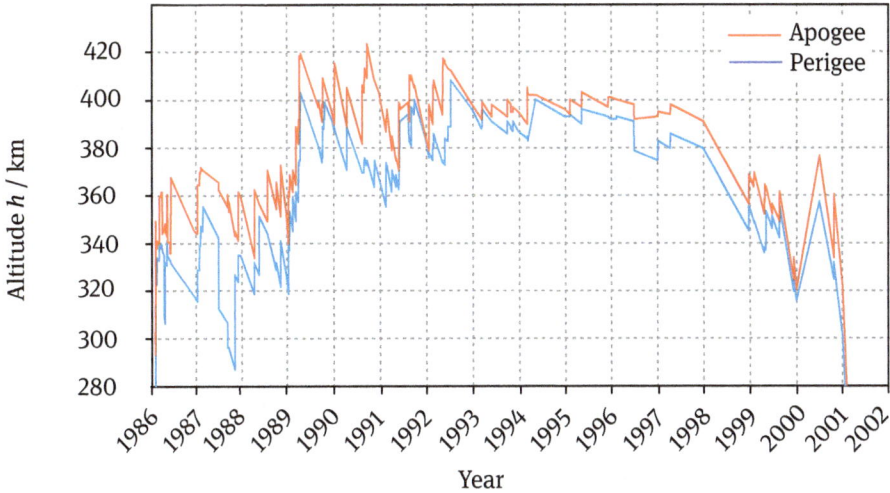

Figure 2.17: Altitude of the Russian Mir Space Station above Earth's surface on an approximated circular orbit. Data: Hall *et al.* [6].

Low Earth orbits tend to downward spiraling in general due to the residual atmosphere of the Earth, always ending with a burn-out in the atmosphere. Therefore, these orbits must be actively lifted up from time to time. The course of the flight altitude of the Russian Mir Space Station during the years of its existence from 1986 to 2001 is shown in Figure 2.17.

A computer generated geometric representation of a LEO trajectory of the International Space Station (ISS) or Mir Space Station is shown in Figure 2.18.

```
1 Orbital plane with Earth      3 Observer's horizon from ground
2 Circular trajectory           4 Observer's horizon from orbit
```

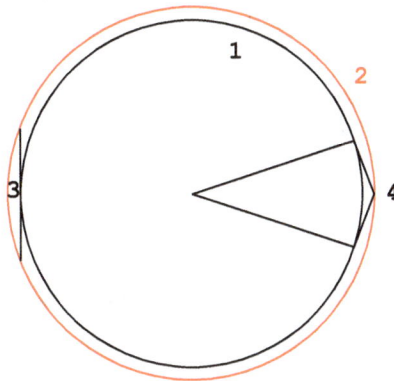

Figure 2.18: Geometrical representation of a typical LEO trajectory (ISS or Mir) at an altitude of approximately 350 km above Earth.

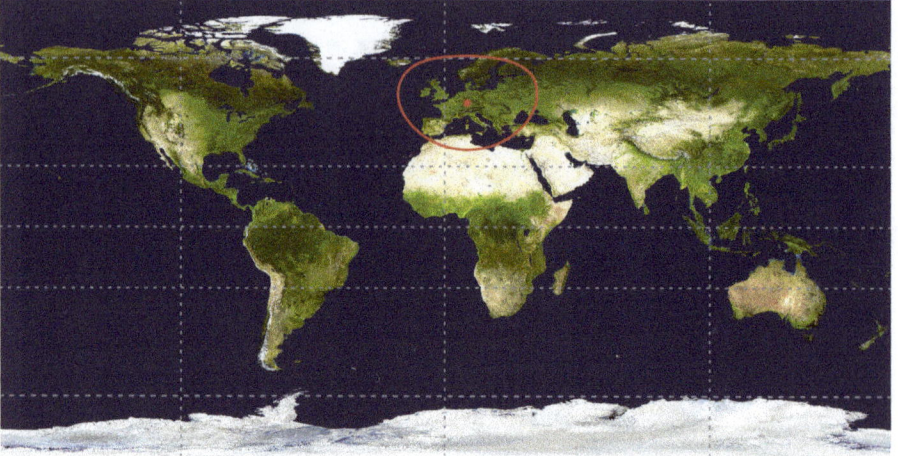

Figure 2.19: Typical visibility range of the ISS when flying over Europe. Credit: NASA [4].

A typical observation range from where crew members on the International Space Station (ISS) photographed the Earth from their point of view is shown in Figure 2.19. The observation range from the station is outlined by a circle and its centre by a red dot.

The strip-shaped fly over, and thus the sine-wave-like ground tracks, are caused by the rotation of the Earth (Figure 2.20). The Earth is rotating under the space station, which itself is moving on a circular orbit around Earth. Computer calculated characteristic quantities of the circular orbit of the International Space Station around Earth are listed below.

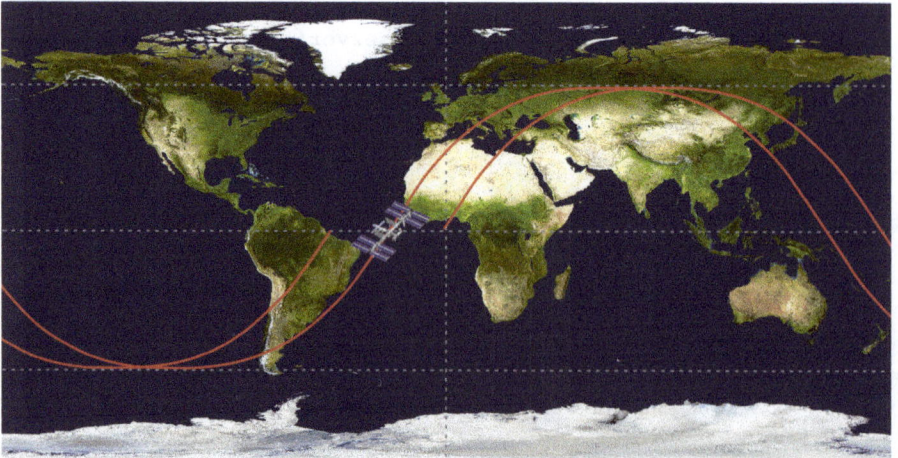

Figure 2.20: Typical ground track of two orbits of the International Space Station. Credit: NASA [4].

Computer generated data of the ground track for Earth

```
Flight time for two orbits :  3h 2m 48s
Altitude (km)               :  350.00
Apse angle                  :  360.00
Latitude                    :  0.00.00
Longitude                   :  - 45.49.27
Velocity (m/s)              :  7700.82
```

Medium Earth Orbit

Mainly the medium Earth orbit (MEO), also called Intermediate Circular Orbit (ICO), at an altitude of 1000 km to 5000 km is of particular importance for future telecommunication applications, and moreover for navigation, geodetic, and space environment sciences. In a global network of satellites, in a so-called constellation, the number of satellites decrease, the larger the viewing area of an individual part of the constellation is. However, due to a lower number of satellites, the running times, limited by the speed of light, of radio communications and optical connections to the satellite and back to Earth are increasing. The existing constellations of Iridium and Globalstar have over 66 or 64 satellites on matched circular orbits at an altitude of approximately 1400 km, respectively.

Geostationary Orbit

The geostationary orbit (GEO) at an altitude of 35 800 km with an inclination of 0 deg is the most frequently used commercial orbit (Figure 2.21). Several hundred satellites already are positioned and other satellites compete for available places in future. Therefore, old used-up satellites were removed from this orbit, if possible, using final fuel supplies and moved to a so-called graveyard, and disposed.

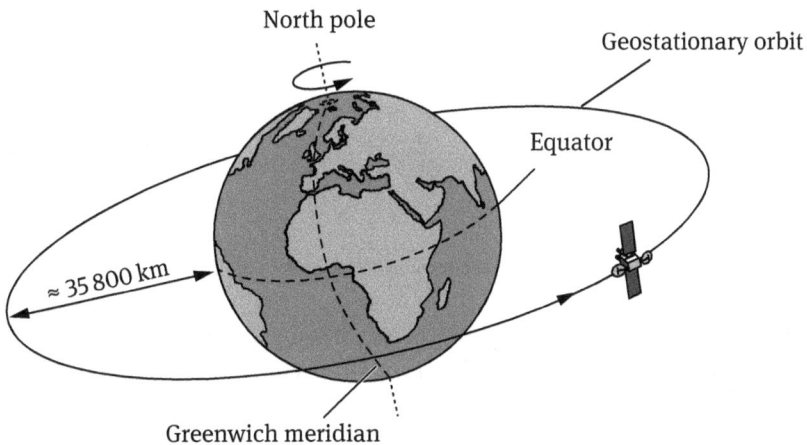

Figure 2.21: Typical geostationary orbit.

```
Altitude   : 3.5800E+04 km                    Tropics and Polar Circles
Longitude  :-3.5000E+01
Latitude   : 0.0000E+00
```

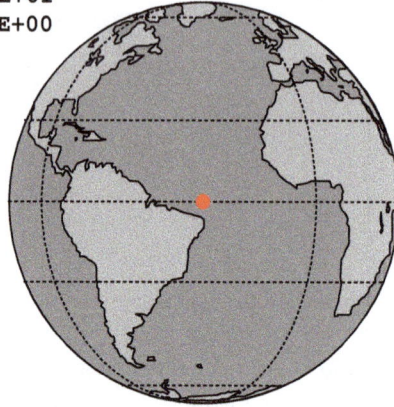

Figure 2.22: Visibility range and subsatellite point (red dot) of a geostationary satellite.

The typical visibility range and the subsatellite point of a geostationary satellite for transatlantic communications are shown in Figure 2.23.

Sun-synchronous Orbit

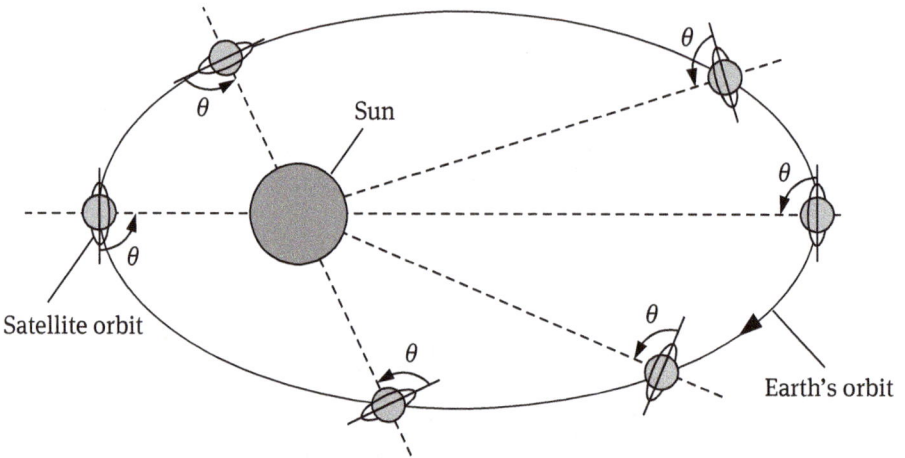

Figure 2.23: Trajectory of an Earth satellite and its position to the Sun. Angle $\theta = 90°$ always.

The sun-synchonous orbit (SSO) at an altitude of 700 km to 3000 km with an inclination of 95° – 100° offers ideal conditions for several Earth observation tasks. Every day, this orbit is moved of less than one degree to right ascension by the influence of Earth's oblateness according to Equation 14.53. A complete revolution of the orbital plane around Earth is achieved during the year thereby.

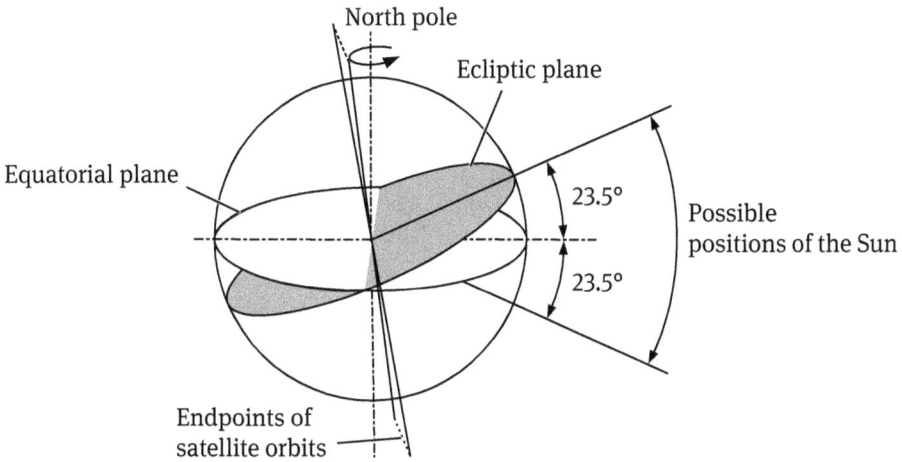

Figure 2.24: Sun-synchronous orbits around the Earth.

Thus, the satellite is synchronised with the Earth's rotation around the Sun in such a way that the satellite never enters the Earth's shadow (see Figure 2.24), and therefore particular demands on energy supply and thermal control are made. The satellite always moves perpendicular to the Sun above the line of daybreak or night fall of the Earth, which naturally cast long shadows by objects on the Earth's surface.

2.5.3 Elliptical Orbit

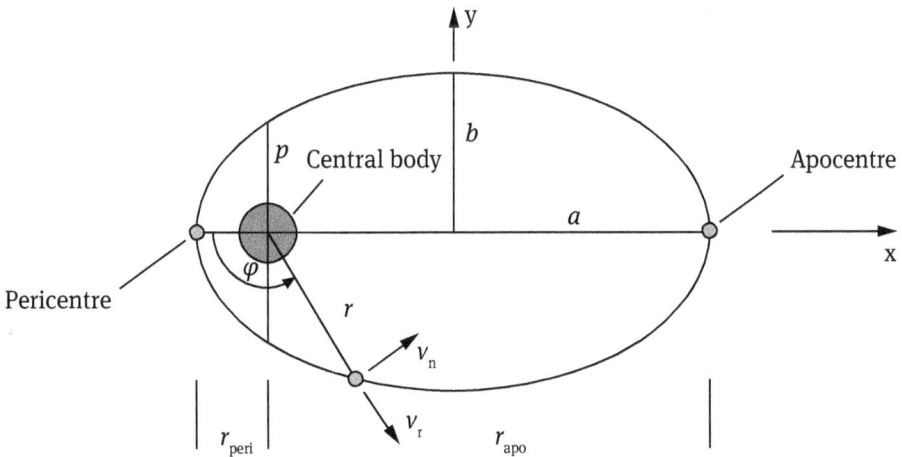

Figure 2.25: Characteristics of an ellipse.

The elliptical orbit is the typical moving of most natural and artificial celestial bodies. This orbit results from an orbit velocity in the pericentre which is greater than the first cosmic velocity and smaller than the second cosmic velocity.

Technical Applications:	Injection of satellites into GEO transfer orbit (GTO)
Geometric reperesentation:	Equations 14.2 to 14.10 (*see* section Geometry, Chapter 14.3, page 198)
Orbit determination:	Equations 14.23 to 14.31, respectively (*see* section Orbit determination, Chapter 14.3, pages 199 to 200)
Limitation:	The mass of the small celestial body is negligible to the mass of the central body.

1 **Orbital plane with Earth**
2 **Position vector to pericentre**
3 **Peri-circular orbit (parking orbit)**
4 **Elliptic trajectory (GTO)**
5 **Apo-circular orbit (GEO)**

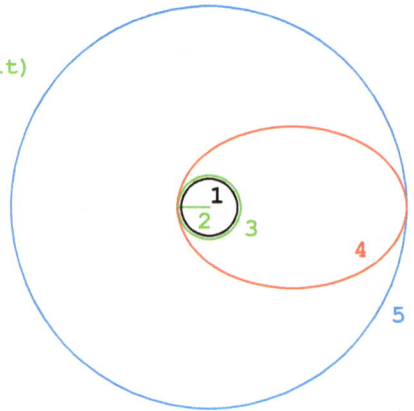

Figure 2.26: Geometrical representation of GTO.

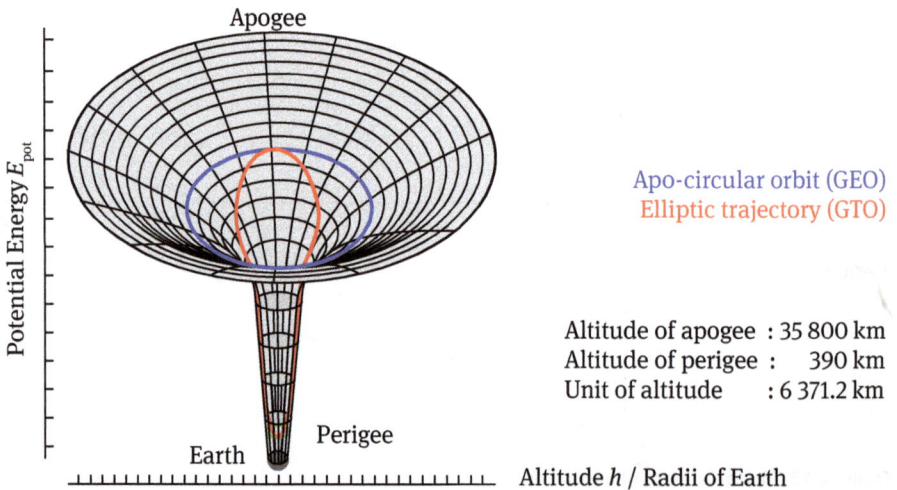

Apo-circular orbit (GEO)
Elliptic trajectory (GTO)

Altitude of apogee : 35 800 km
Altitude of perigee : 390 km
Unit of altitude : 6 371.2 km

Figure 2.27: Spatial visualisation of a GEO transfer orbit (GTO).

Trajectory within Earth's gravity field

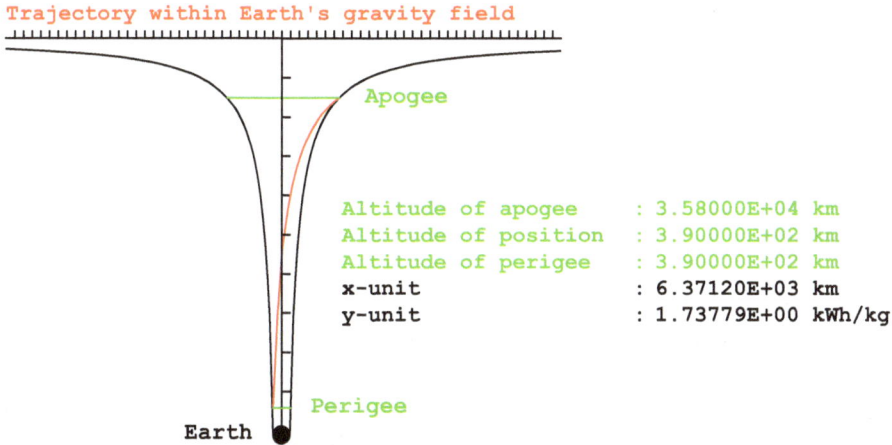

Apogee

Altitude of apogee	: 3.58000E+04 km
Altitude of position	: 3.90000E+02 km
Altitude of perigee	: 3.90000E+02 km
x-unit	: 6.37120E+03 km
y-unit	: 1.73779E+00 kWh/kg

Perigee

Earth

Figure 2.28: Representation of a GTO in the gravitational field of the Earth.

Viewed from the side, the red elliptical trajectory no. 4 (GTO) from Figure 2.26 is shown in Figure 2.28. On the ellipse the perigee is the closest point to the Earth and the apogee is the furthest point to the Earth. A spatial, three-dimensional representation of GTO and GEO is shown is Figure 2.27. Figure 2.28 illustrates the computer calculated two-dimentional course of the elliptical trajectory no. 4 from Figure 2.26 within the gravitational field of the Earth. The ground track of the GTO is visualised in Figure 2.29. The 's-shaped' ground track is due to Earth's rotation under the satellite while the satellite itself is moving on an elliptic trajectory around Earth. The computer calculated characteristics of the GTO are listed below.

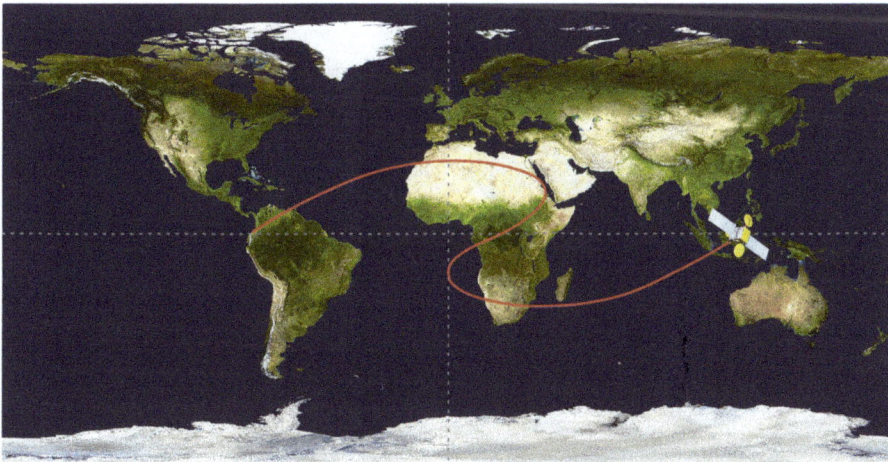

Figure 2.29: Ground track of a GEO transfer orbit (GTO). Credit: NASA [4].

Computer generated data of the ground track for Earth

```
Flight time              : 10h 34m 46s
Altitude (km)            : 390.00
Apse angle (deg)         : 360.00
Latitude                 : 0.00.00
Longitude                : 120.22.20
Velocity (m/s)           : 10080.30
```

The GEO transfer Orbit (GTO) ranges from the perigee, the closest distance to Earth, to the apogee, the furthest distance to Earth, at the height of the geo-stationary orbit. After end of firing of the launcher system close to the Earth, the satellite is moving powerlessly to its highest orbit point according to the laws of orbital mechanics. At this point the velocity has to be increased by one or more propulsion manoeuvres to reach the necessary orbital velocity allowing a lasting positioning above the equator. Currently, several hundred satellites in the GEO actively provide users with numerous services for weather observation and communication within their range of visibility.

Molniya Orbit

Satellites of the Russian Molniya series move on a particularly stable orbit. They use the influence of Earth's oblateness on the orbital element 'Argument of Perigee' according to Equation 14.54. As a result, while at an orbital inclination of approximately 63.5°, the perigee of the orbit is stably placed above the southern hemisphere and therefore, a high percentage of availability above the Russian territory is realised. Computer calculated characteristics of the orbit are listed below.

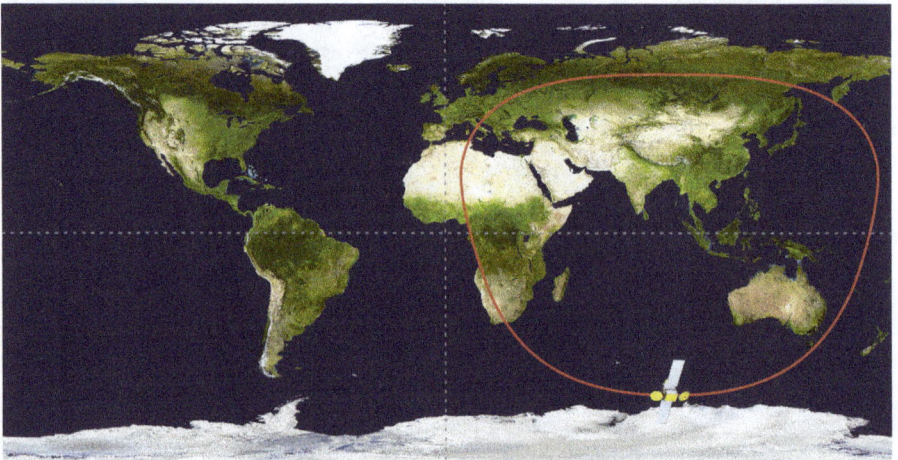

Figure 2.30: Ground track of a 24h-Molniya trajectory. Credit: NASA [4].

Computer generated data of the ground track for Earth

```
Flight time          :   23h 56m 4s
Altitude (km)        :   2061.52
Apse angle (deg)     :   360.00
Latitude             :   -63.29.59
Longitude            :   89.59.59
Velocity (m/s)       :   9223.87
```

A Molniya orbit is well suited to communications in the region shown in the above Figure 2.30. The advantage of such an orbit is that considerably less launch energy is needed to place a satellite into a Molniya orbit than into a geostationaty orbit. There are two disadvantages of such an orbit. The ground station needs a steerable antenna to track a satellite on the this orbit. Another disadvantage is that the satellite passes the Van Allen radiation belt four times a day. The Van Allen belt contains high energetic electrons and protons, from which the satellite needs to be protected.

After the end of the Cold War the use of the Molniya orbit more and more decreased. In the former Soviet Union this orbit was of essential importance for military use. In the west there were no significant applications.

2.5.4 Parabolic Trajectories

The parabolic trajectory is the mathematical solution of the transfer from a returning (elliptical) orbit to a non-returning (hyperbolic) trajectory with the following boundary conditions.

- Eccentricity of conic section $e = 1$
- Semi-major axis $a -> \infty$
- Period of orbit $T -> \infty$

In nature such mathematically accurate trajectories do not occur. However, this representation is often applied as approximated description. This type of trajectory results from a trajectory velocity at the pericentre which is equal to the second cosmic velocity. Sometimes the cosmic velocities were also referred as astronautical velocities during the 'Space Race' between the United States and the Soviet Union in the 1950s and 1960s.

Geometrical representation: Equations 14.11 to 14.13 (*see* section Geometry, Chapter 14.3, page 198)

Calculation of trajectory: Equations 14.23 to 14.29, and 14.32 with $e = 1$ and $a -> \infty$, respectively (*see* section Orbit determination, Chapter 14.3, pages 199 to 200)

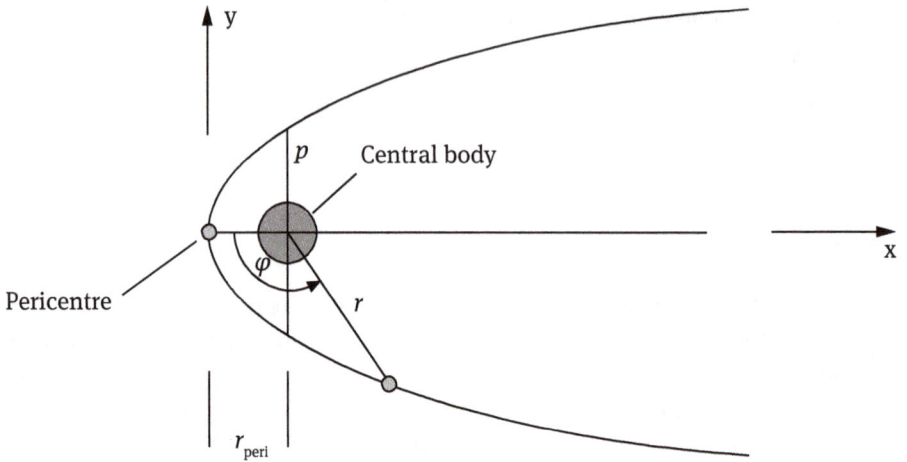

Figure 2.31: Characteristics of a parabola.

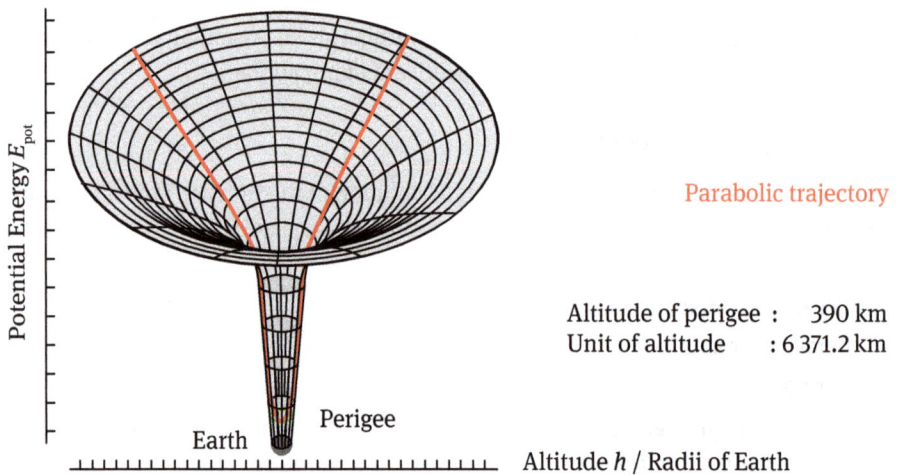

Parabolic trajectory

Altitude of perigee :   390 km
Unit of altitude   : 6 371.2 km

Altitude h / Radii of Earth

Figure 2.32: Geometical representation of a parabolic trajectory around the Earth (not to scale).

At the perigee the celestial body has exactly the second cosmic velocity, while far away from Earth, it has no residual velocity anymore (Figure 2.32). To reach other celestial bodies out of the gravitational field of the Earth, space probes have to be accelerated to a velocity above the second cosmic velocity concerning the height of the perigee. In 1962, this minimum requirement was exceeded for the first time during the Mariner 2 mission. Currently, less than five space probe leave the gravitational field of the Earth each year, because of high technical requirements and significant financial costs.

Trajectory within Earth's gravity field

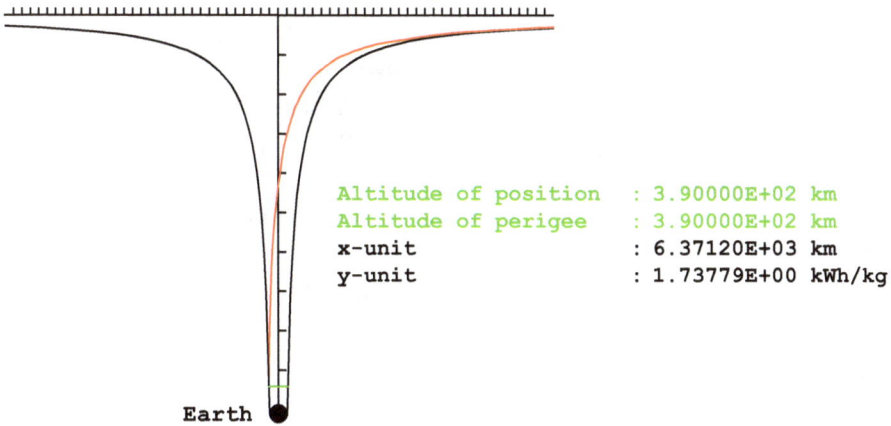

Altitude of position	: 3.90000E+02 km
Altitude of perigee	: 3.90000E+02 km
x-unit	: 6.37120E+03 km
y-unit	: 1.73779E+00 kWh/kg

Earth

Figure 2.33: Two-dimensional representaton of a parabolic trajectory in Earth's gravitational field.

2.5.5 Hyperbolic Trajectories

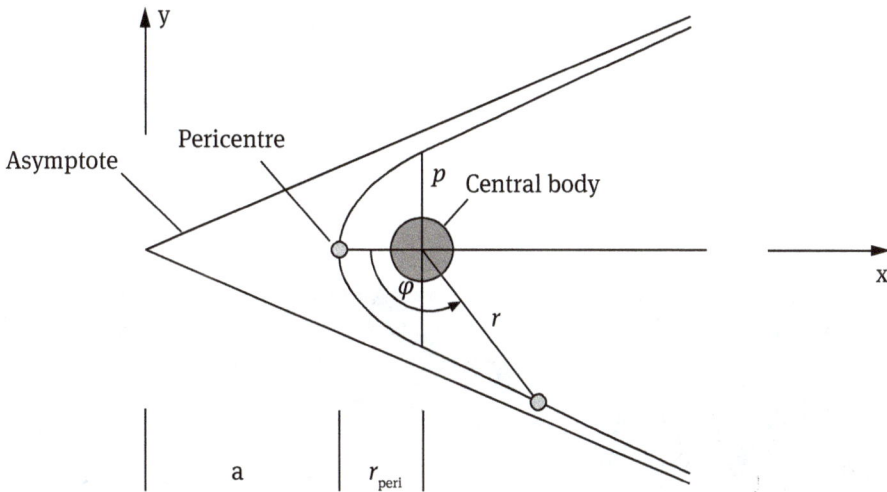

Figure 2.34: Characteristics of a hyperbola.

The hyperbolic trajectory is the typical form of the motion of all natural and man-made celestial bodies, which are not periodically return closely to a mass body. This type of trajectory is the result of a trajectory velocity at the pericentre which is greater than the second cosmic velocity.

Technical applications:	Fly-by of celestial bodies for exploration.
	Fly-by of celestial bodies for changing of orbit.
	(swing-by manoeuvre, see Chapter 2.3.6)
Geometrical representation:	Equations 14.14 to 14.22 (*see* section Geometry,
	Chapter 14.3, page 199)
Calculation of trajectory:	Equations 14.23 to 14.29, and 14.33 (*see* section
	Orbit determination, Chapter 14.3, pages 199 to 200)
Limitation:	Mass of a small celestial body is negligible
	to the mass of the central body.

1 Orbital plane with Earth
2 Position vector with position orbit
3 Hyperbolic orbit
4 Peri-circular orbit
5 Trajectory asymptotes

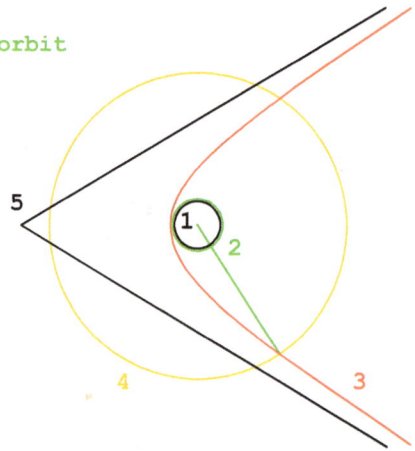

Figure 2.35: Geometrical representation of a hyperbolic trajectory around Earth ($v_\infty = 3\,000\ \mathrm{m \cdot s^{-1}}$).

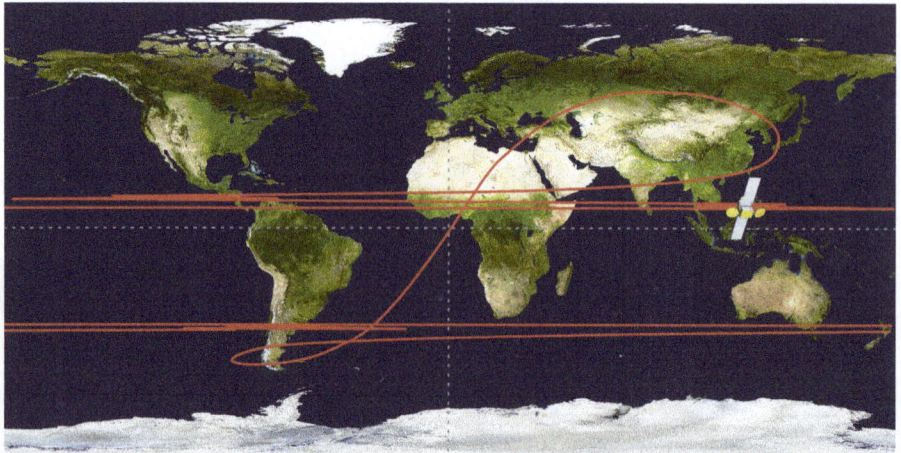

Figure 2.36: Ground track of a hyperbolic trajectory. Credit: NASA [4].

Computer generated data of the ground track for Earth

```
Flight time          :  4w 3d 1h 34m 39s
Altitude (km)        :  8257955.50
Apse angle (deg)     :  150.00
Latitude             :  8.10.39
Longitude            :  119.55.49
Velocity (m/s)       :  3016.03
```

The characteristic ground track of a hyperbolic trajectory with $v_\infty = 3000 \, \text{m} \cdot \text{s}^{-1}$ and an inclination of trajectory of 55° with perigee above Africa is shown in Figure 2.36. The ground track is incomplete at the southern approaching latitude and at the departing northern latitude.

2.5.6 Trajectory Changes

Forces are necessary to change the state of motion of a natural or man-made celestial body. Natural bodies can only exert forces by their mass action or by a decelerating atmosphere. Technical applications of natural trajectory changes are gravity-assist manoeuvres, swing-by, and the so-called aerobraking which are described in the sections 'Gravitational Manoeuvres' and 'Atmospheric Braking Manoeuvres' further below. Man-made celestial bodies (spacecrafts) are able to exert additional forces specifically by their own propulsion system. In the subsequent chapters today's predominantly applied procedures for desired trajectory changes are outlined.

The deceleration of low-flying earth satellites by residual atmosphere results in downward spiraling into more dense atmospheric layers with increasing braking action (see section 'Spiral Transfers' below). To avoid a crash with burn-out in Earth's atmosphere, this unwanted troublesome trajectory change must be compensated from time to time by complex boosting manoeuvres. Multiple boosting manoeuvres of the Russian Mir Space Station during its existence are illustrated in Figure 2.17.

Hohmann Transfer Orbits

In 1925, long before one might think of technical implementation, *Walter Hohmann* investigated the transfer between two orbits scientifically. He showed that for an energetic favourable transfer between two orbits with radius r_1 and radius r_2, with $r_2/r_1 < 12$, the use of one semi-elliptical transfer orbit is most favourable (see Equations 14.40 to 14.42. For $r_2/r_1 > 12$ the 3-impulse transfer with two semi-elliptical transfer orbits according to Equation 14.43 is most favourable.

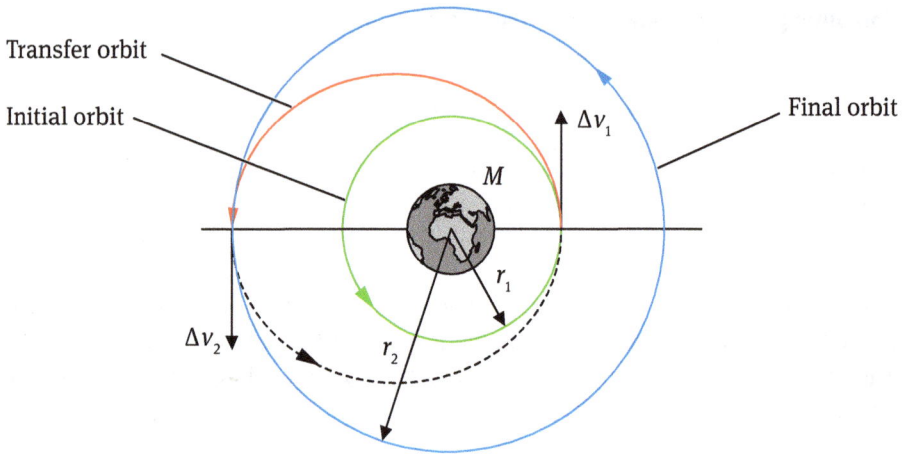

Figure 2.37: Hohmann-2-impulse transfer.

During 2-impulse transfer the period of transfer results from the half of the orbital period on the elliptical orbit according to Equation 14.30. Assuming $\Delta v_2 = 0$ during 3-impulse transfer, this orbital impulse must be performed mathematically in an infinite distance from the central body. The time of flight of the two semi-ellipses also tend to infinity. Therefore, such transfers are of a more theoretical nature.

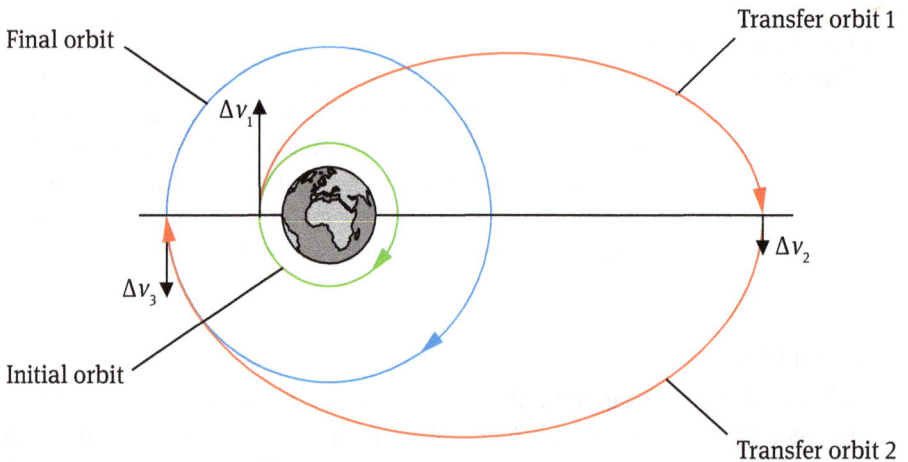

Figure 2.38: Hohmann-3-impulse transfer.

For technical implementation propulsion manoeuvres are often performed in multiple small steps. Thus, a more accurate injection into the desired destination orbit is possible and unnecessary propellant consumption within the destination orbit for fine adjustment is avoided.

If two orbits have an infrastructure independently from each other (e. g., a space station within LEO and satellite within GEO) or natural celestial bodies (e. g., Earth and Venus), the launching time must be scheduled so that on approaching the destination orbit the desired celestial body is nearly reached, too. The approach of two celestial bodies is called 'rendezvous'. For this, so-called launch windows must be timed. These launch windows repeatedly occur according to the synodic period (see Equation 14.37).

Spiral Transfers

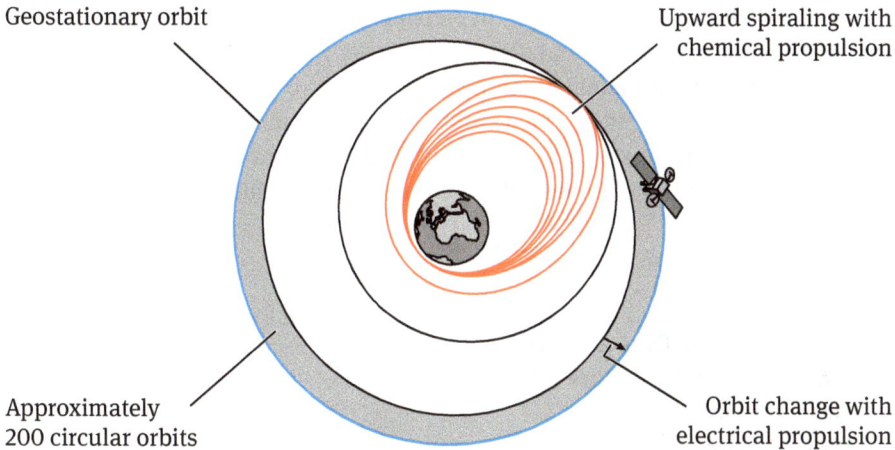

Geostationary orbit

Upward spiraling with chemical propulsion

Approximately 200 circular orbits

Orbit change with electrical propulsion

Figure 2.39: Orbit ascent of Artemis.

Alternatively, flight from a launch orbit to a destination orbit can also be operated by spiralling. The required velocity for this transfer is greater than of the 2-impulse transfer (see Equation 14.44). For these transfers low-thrusted engines were used for continuous operation. Also solar sailing is a method of propulsion technology for such orbit transfers. The period of transfer may constitute up to several years, and it is supposed that a man-made celestial body is continuously moving with orbiting data of a quasi circular orbit whose radius is continuously increasing and decreasing with downwarding orbit, respectively. The increased propellant consumption for this orbit transfer can also be explained by lifting up a bigger mass within the gravitational field of the central body. Therefore, potential energy is 'consumed'. Propulsion manoeuvres are most effective (e. g. gain of velocity) if propellant is ejected for this as 'deep' as possible within the gravitational field of the central body.

In 2001, a failure occurred in the propulsion system of the upper stage of the Ariane 5 rocket during launch of the European technology satellite Artemis (Advanced Relay and Technology Mission). Therefore, the satellite was launched into a lower orbit. As shown in Figure 2.39 a combination of boosting manoeuvres with the help of the chemical propulsion system up to an altitude of 31 000 km (inner boundary of the grey ring) and subsequent help of the electrical propulsion system, the satellite succeeded in reaching its GEO destination orbit by its own means at an altitude of 35 800 km (outer boundary of the grey ring). The increased propellant consumption during ascent results in a shorter lifetime of the satellite than originally planned. The upward spiraling took several months to reach GEO.

Gravitational Manoeuvres

For a geometrical representation of a gravitational manoeuvre (swing-by gravity assist), we take the drawing of the hyperbolic trajectory from Chapter 2.3.5 and adding the approach velocity and departure velocity v_∞. The Law of Conservation of Energy defines that the amount of the velocity vector v_∞ is the same at approach and at departure far away from the central body with a deviation angle Φ (see Equation 14.39). An unaffected swing-by distance is calculated according to Equation 14.38. The Law of Conservation of Angular Momentum defines that

$$n \cdot v_\infty = r_{peri} \cdot v_{peri} \tag{2.16}$$

Where
n is the unaffected flyby altitude[3]
v_∞ is the approach / departure velocity
r_{peri} is the distance of perigee
v_{peri} is the velocity at the perigee

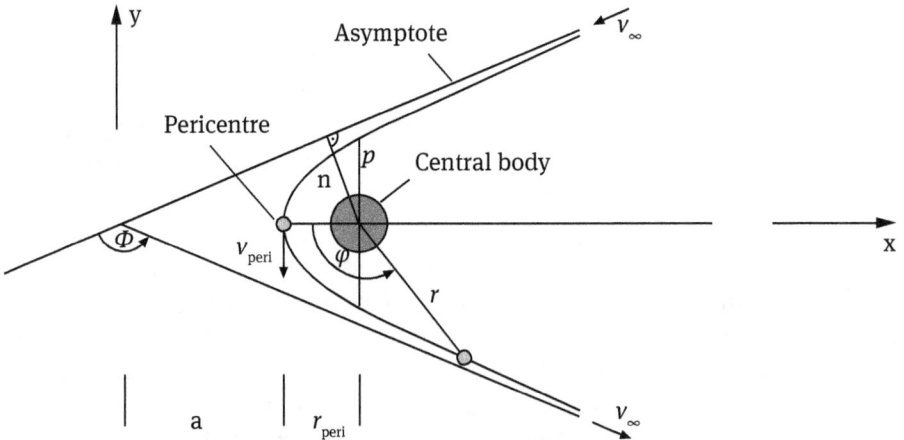

Figure 2.40: Geometrical representation of a gravitational manoeuvre.

---

3 No deviation of the approaching body occurs, if the central body has a very small mass.

Here, we consider the pure (i.e. passive) gravitational manoeuvre. If propulsion manoeuvres are performed 'deep' within the gravitational field during swing-by of a heavy celestial body, the effect of the gravitational manoeuvre is considerably increased.

Regarding the next superordinate central body (e.g., the Sun at approach of a planet or planet at approach at one of its moons) the maximum gain of velocity is equivalent to the orbital velocity of the altitude of the pericentre r_{peri}. Thus, a gain of velocity is realised without fuel consumption of up to $7 \, km \cdot s^{-1}$ for Earth or Venus and between $15 \, km \cdot s^{-1}$ and $40 \, km \cdot s^{-1}$ for the gas planets, Jupiter, Saturn, Uranus, and Neptune. Regarding the overall system the amount of energy must be constant, e.g. consequently, at an acceleration of the satellite a decelerating of the central body approached takes place on its orbit around the next superordinated centre of gravity or *vice versa* during deceleration of the satellite.

A computer calculated geometrical representation of the velocity vectors during a gravitational manoeuvre for a plane approach at Jupiter (e.g., the satellite is approaching on the trajectory plane of the planet and also escapes from the same plane) is illustrated in Figure 2.41.

The gravitational manoeuvre is determined by the velocity vectors of the planet, of the satellite at approach, and by the deflection angle (Equation 14.39). The circle is formed by the velocity vector v_{∞} and the centre of the circle at the end of the velocity vector of the planet. The endpoints of the velocity vectors of the satellite are always on the circular line. The types of orbits which result from these velocity vectors in Figure 2.41, before and after the gravitational manoeuvre, are presented in Figure 2.42. The circular orbit and the location vector of planet Jupiter and the orbital changes of the satellite are illustrated. The orbital shape afterwards is projected into the orbital plane before (Figure 2.42).

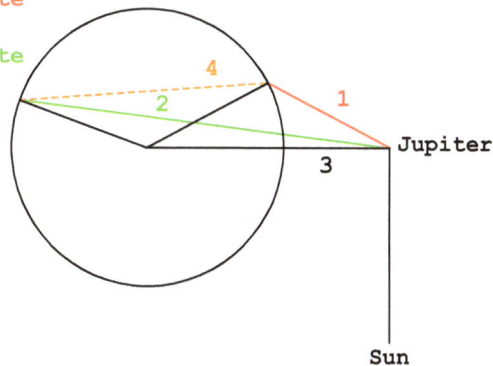

x,y-plane of Jupiter and satellite before swing-by
(z: vertical to x,y)
1 Velocity vector of satellite before swing-by
2 Velocity vector of satellite after swing-by (ends on a surface of a sphere)
3 Velocity vector of Jupiter
4 Δv-kick

7.4805E+03 m/s
2.0255E+04 m/s
1.3176E+04 m/s
1.3473E+04 m/s

Figure 2.41: Geometrical representation of velocity vectors during a gravitational manoeuvre.

```
Before swing-by
Perigee height   : 1.1433E+08 km
Eccentricity     : 7.5295E-01
Apogee height    : 8.1549E+08 km
Apse angle(deg)  :      169.76
Location tangent :      117.33
```

```
After swing-by
Perigee height   : 7.6696E+08 km
Eccentricity     : 1.4007E+00
v-infinity       : 8.3232E+03 km
Apse angle(deg)  :       12.49
Location tangent :       97.29
```

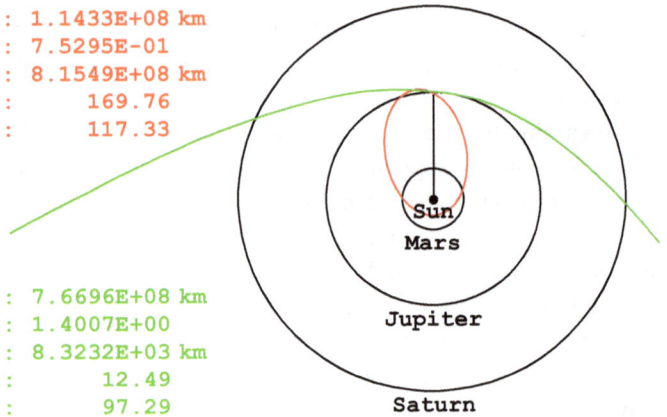

Figure 2.42: Types of orbits, before and after gravitational manoeuvre.

In the example mentioned above an elliptical orbit around the Sun changes to a hyperbolic orbit with a residual velocity exceeding $8\,\text{km}\cdot\text{s}^{-1}$ without propellant. At a deflection angle of approximately 130° a manoeuvre to change the orbit can be accomplished within approximately 10 h duration of stay nearby (satellite within apse altitude of $-90° < \varphi < +90°$) that practically cannot be realised with chemical propulsion systems.

For a complete and more general representation with a launch angle outside of the trajectory plane of the planet, the planar circle changes to a spatial sphere with radius v_∞, as it can be seen from the previous two-dimensional velocity vector diagram (see Figure 2.43). Any other relations shall apply *mutatis mutandis*.

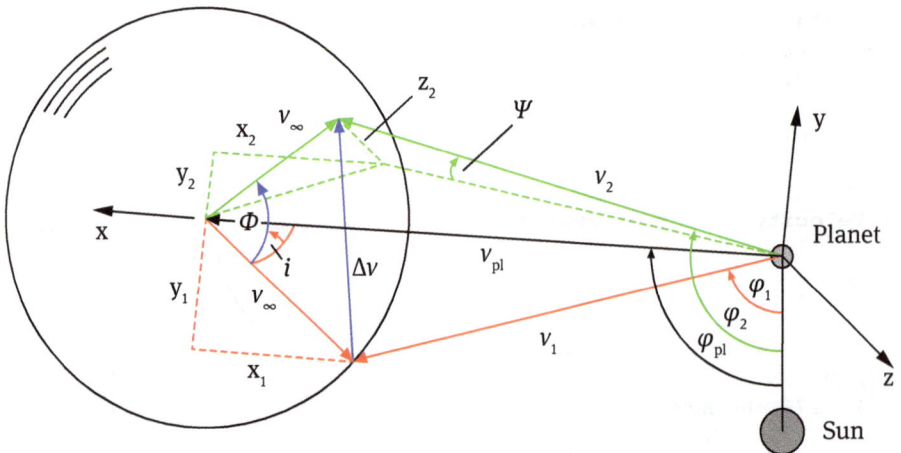

Figure 2.43: General representation of the velocity vector diagram (not to scale).

The angle of inclination i is the angle at which the satellite approaches the planet with respect to the approach plane x,y (approach latitude, however, not to be confused with the latitude of the planet which refers to the rotation of the planet). By choosing the design of the approach trajectories (φ_1, v_1, i) carefully, celestial destinations can be reached which would be otherwise inaccessible with currently available technology.

Such a manoeuvre took place during the Ulysses mission in 1992. To achieve a high inclination of 80°, flyby at Jupiter was necessary. Coming from Earth's plane of ecliptic the probe was catapulted into a steeply sloping elliptic orbit around the Sun. This was the only way to realise the requirements for investigating the polar areas of the Sun. Basically, such swing-by manoeuvres may be performed several times at the same celestial body. Before successfully injecting into orbit around planet Mercury, the first Mercury space probe was lauchend in 2004 and reached its destination orbit only after one flyby at Earth, two flybys at Venus, and three flybys at Mercury.

Examples

- With small v_2 the pericentre of the orbit around the superordinate central body (here the Sun) can be reduced and, in extreme cases, to fly into the Sun (not possible directly from Earth).

- With large departure angle of inclination Ψ destinations far outside of the ecliptic (Earth's orbital plane around the Sun) can be reached.

- Celestial bodies are more reachable on very different orbits (comets, Pluto, and Mercury) by suitable gravitational manoeuvres.

2.5.7 Atmospheric Braking Manoeuvre

The possibility of aerobraking is another procedure of an orbital manoeuvre without propellant consumption at some celestial bodies which have an atmosphere. Here, an aerodynamic drag F is realised by friction in the high atmosphere.

$$F = \frac{1}{2} c_W A_{eff}\, \rho\, v^2 \tag{2.17}$$

Where

F is the aerodynamic drag
c_W is the drag coefficient
A_{eff} is the effective cross-sectional area
ρ is the density of the atmosphere
$\vec{v}$ is the velocity

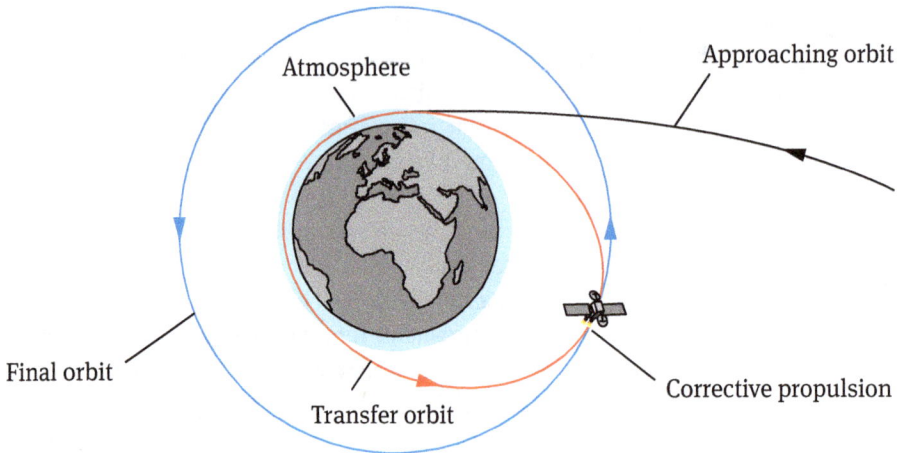

Figure 2.44: Touching the atmosphere to provide a new satellite orbit.

Aerobraking requires precise knowledges of the composition of the atmosphere at the time of braking (for c_w and ρ). The density of the high atmosphere may vary considerably. There is a significant risk of total loss if the strictly defined entry corridor is not reached. If the entry angle is too large a significant additional thermal load may occur which can lead to burn-out. If the entry angle is too flat there is a risk of 'bouncing off' like a stone which is thrown in parallel to the surface of water with high velocity.

In contrast to gravity manoeuvres only deceleration but no acceleration of the satellite can be accomplished during aerobraking. Therefore, this procedure is useful on arrival of an interplanetary satellite at a destination body (e.g. Mars or Saturn's moon Titan) as alternative to propellant consuming braking manoeuvres in the pericentre. However, a propulsion system is indispensible because in the apocentre of the injection orbit an acceleration manoeuvre has to proceed to avoid repeated entry into the atmosphere with inevitable burn-out or crash. Also for future return flight from GEO of the Earth into LEO (e.g. with ISS rendezvous) such a procedure could only offer the necessary economic basis.

2.5.8 Multiple-body Problems

In the previous chapters we used the two-body problem as a basis. This constituted an approximation to reality which was sufficient for our considerations. For technical implementation in space flight applications we have to consider interferences and deviations.

We assumed that a small body is moving around a large body provided that $m \ll M$. This is applied to all man-made celestial bodies with sufficient accuracy.

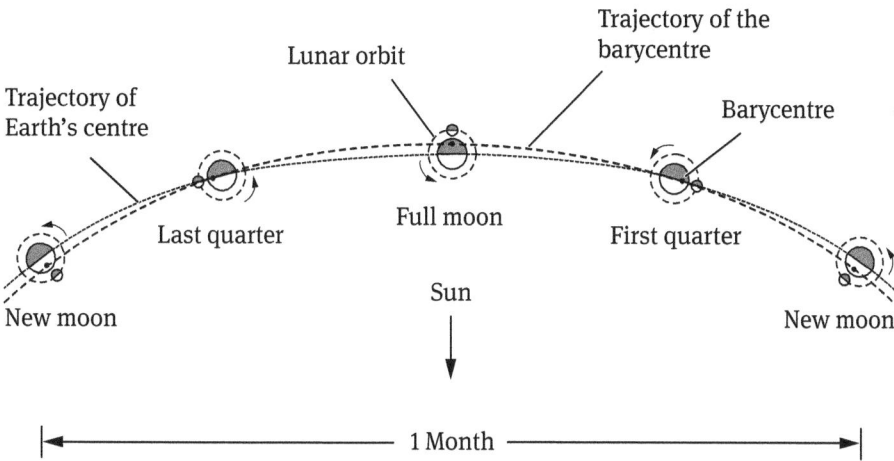

Figure 2.45: Rotation of the Earth and the Moon around the common barycentre.

On natural planets and moons this approximation cannot be applied anymore without restriction. Therefore, we have to take into account that the motion of two celestial bodies occurs around a common centre of gravity, the barycentre.

Furthermore, we expected that masses of celestial bodies are always distributed homogeneously within a sphere with constant radius R_0. All of this does not occur in reality. Every celestial body exhibits a so-called oblateness due to its rotation, i.e. the distance between the north and south poles is smaller than the diameter of the equator. This has an impact on the orbital elements of satellite's orbits and is used in part to reach certain destination orbits (SSO, Molniya orbits). By means of inhomogeneous mass distribution the gravitational force of celestial bodies also depends from the geographical location. Satellites move on orbits which do not meet exactly the conic sections mentioned in the previous chapters.

We also expected that only two celestial bodies are regarded (e.g. satellite around Earth). In reality all masses attract each other and therefore, move around their own common barycentre. In some special cases there are even analytical solutions for three-body problems (see section 'Lagrangian Points' below). However, calculations in celestial mechanics must be performed with numerical methods in general, for which there are very precise solutions by high computing power today. For satellite orbits around Earth the impacts of the Sun and Moon have at least taken into account. In particular they affect the stationing of GEO satellites where these interferences (approximately $150 \, \text{m} \cdot \text{s}^{-1}$ per year for corrections from south to north among others) must be corrected regularly by the propulsion system. Propellant is consumed in the course of this. Today, the availability of propellant limits the lifetime of these satellites.

Given by the spatial conditions in our astronomical environment every body reaches certain limits as the distance to the central body increases where the present interferences from outside become greater compared with the impact of the central body as such. It can be clearly said that if we are regarding a gravitational field around the Earth for the GTO as mentioned in Chapter 2.3.3, the Moon and all other celestial bodies, too have a gravitational field (with varying 'depth') around. So, this finding provides better insights when and which a celestial body must be identified as central body. There is a number of other aspects which should not be described in detail here, including

- forces by magnetic influences,
- forces by existing residual atmosphere, and
- forces by radiation pressure.

Neutrosphere

Every mass body has a gravitational field with a limited sphere of influence. The simplifications made in the previous chapters, in particular the assumption of a two-body problem, are only accepted within the so-called neutrosphere of the celestial body (e.g. the Earth). Beyond this sphere of influence the effects of the superordinate central body (e.g. the Sun) should be considered. The ceiling of the neutrosphere opposite to the superordinate centre of gravity can be calculated from

$$\left(\frac{r_n}{r_s}\right)^2 = \frac{m_n}{m_s} \tag{2.18}$$

Where

r_n is the radius of neutrosphere
r_s is the distance from superordinate centre of gravity
m_n is the mass of celestial body
m_s is the mass of superordinate centre of gravity

The Earth's neutrosphere is of approximately 260 000 km compared with the Sun. The Earth's Moon moves beyond the neutrosphere and it is attracted stronger by the Sun than by Earth as central body. Since Earth is rather attracted (accelerated) equally by the Sun, the differential accelerations are crucial allowing a stable orbit of the Moon around Earth.

Activesphere

If we depart from Earth there must be a distance limit, from which a satellite moves away from its orbit around the Earth and does not return anymore. The satellite then follows an Earth-like orbit around the Sun. The ceiling of this sphere of influence is referred to as activesphere. Beyond this activesphere no stable orbits are possible around the central body. The ceiling of the activesphere opposite to the super–ordinate centre of gravity can be calculated from

$$\left(\frac{r_n}{r_s}\right)^5 = \frac{1}{2}\left(\frac{m_n}{m_s}\right)^2 \tag{2.19}$$

Where

r_n is the radius of the activesphere
r_s is the distance to the superordinate centre of gravity
m_s is the mass of the celestial body
m_n is the mass of the superordinate centre of gravity

This results in an activesphere of approximately 800 000 km for the Earth compared with the Sun. The Earth's Moon moves within this activesphere. However, there are some more positions beyond this activesphere which are interesting for space applications (Lagrangian points).

Lagrangian Points

The Lagrangian points or libration points are the most distant positions from the central body for which there are, in terms of figures, permanently applicable positions of satellites. It refers to five special solutions of the three-body problem. The situation is clearly shown in Figure 2.46 for the points L_1 and L_4.

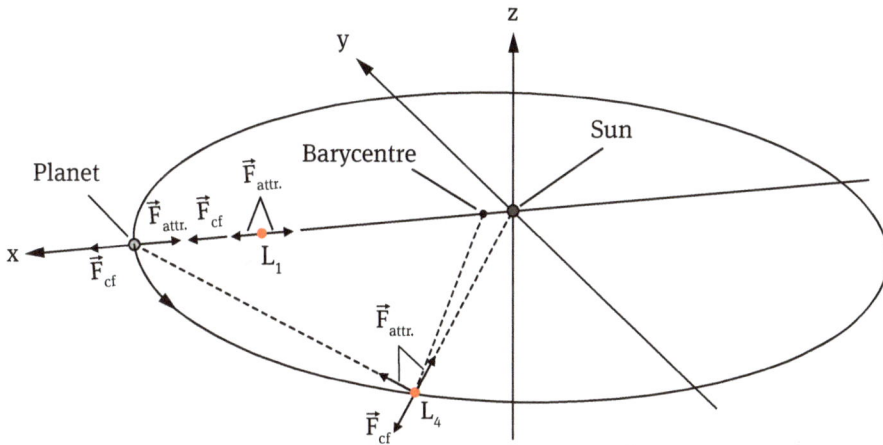

Figure 2.46: Equilibrium in the Lagrangian points L_1 or L_4 (not to scale).

Where

$F_{attr.}$ is the force of attraction
F_{cf} is the centrifugal force
L_1, L_4 are Langrangian points

The distribution of the five Lagrangian points for the systems Sun-Earth and Earth-Moon are shown in Figures 2.47 and 2.48 below.

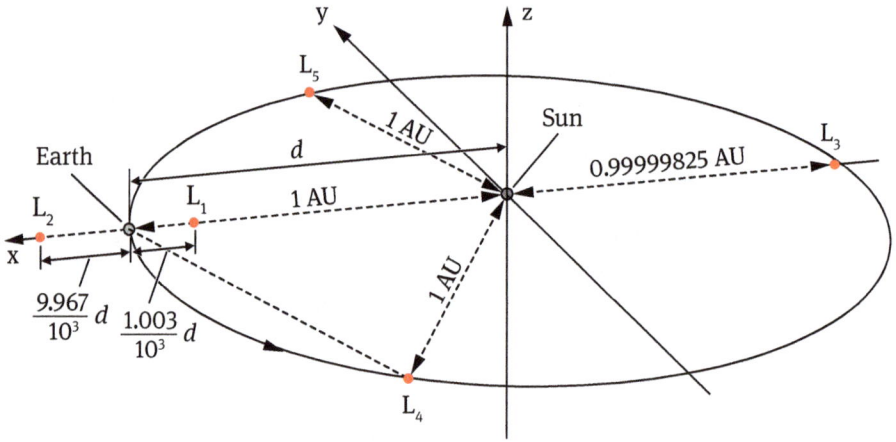

Figure 2.47: System Sun-Earth (not to scale).

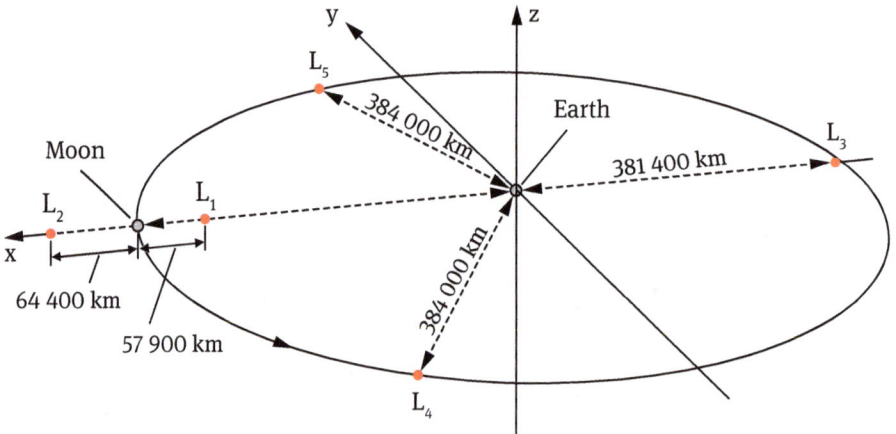

Figure 2.48: System Earth-Moon (not to scale).

In terms of figures there is an equilibrium of forces at the five L-points for the gravitational force and the centrifugal force in such a way that the geometric distances remain constant. It can be proven that the points L_1 to L_3 are unstable and the points L_4 and L_5 are stable. Small interferences which are always present thus cause a 'shift' from L_1 to L_3, while L_4 and L_5 are occupied by natural small planets in a club-like area, e. g., at Jupiter.

The Solar and Heliospheric Observatory (SOHO) satellite was placed in a curly orbit around L_1 to observe the Sun and to early warning. Thus, it is always located approx. 1 500 000 km from Earth, optically in front of the Sun. The satellite is able to send instant data of solar wind and particle stream to Earth with a radio time delay of approximately 5 s.

Interplanetary Flights

Leaving the sphere of influence of the Earth a satellite takes along the original motion state of the Earth as central body, e.g., the velocity vector of the Earth $v_\infty \approx 30 \, \text{km} \cdot \text{s}^{-1}$ has to be added to the velocity vector v_0 of the satellite by direction and amount. The below Figure 2.49 illustrates this representation.

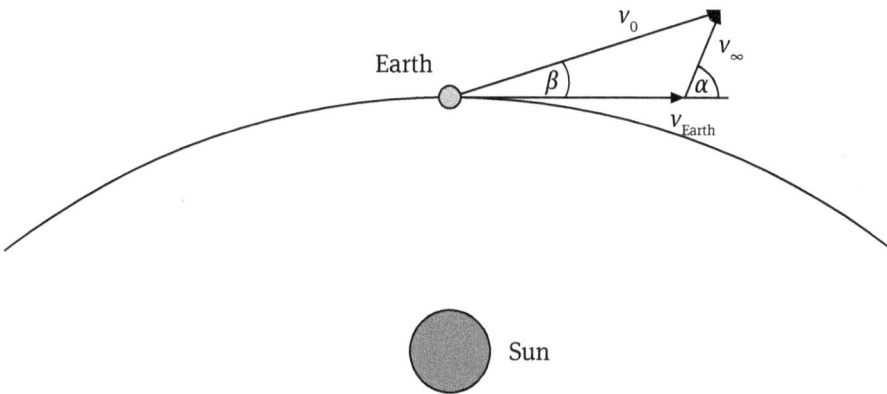

Figure 2.49: Addition of velocity vectors.

The satellite then moves around the Sun as independent celestial body with a velocity v_0. In addition, if the satellite should also leave the solar system directly from Earth (e.g., on a parabolic orbit) the satellite is controlled in such a way that the angle becomes $\alpha = 0°$ and the velocity vector v_0 corresponds to the velocity vector vperi of the parabolic orbit around the Sun (here approximately $42 \, \text{km} \cdot \text{s}^{-1}$). A velocity v_∞ of approximately $12.4 \, \text{km} \cdot \text{s}^{-1}$ is required during lift-off from Earth's surface ($h = 0$) to reach this velocity. The velocity v_∞ is reached according Equation 14.24 at a lift-off velocity vperi of approximately $16.7 \, \text{km} \cdot \text{s}^{-1}$ from Earth's surface. Thereby, the satellite reaches a galactic orbit independent from the Sun. In literature this velocity is often designated to as third cosmic velocity. According to that a fourth galactic velocity can be defined for lift-off from Earth's Moon into a galactic orbit in the Milky Way, etc.

To assess the feasibility of an interplanetary flight, it is also important which power requirement is needed for an approach to and a departure from this celestial body. The velocity requirement for a permanent orbit around the target body is limited in its maximum amount by the requirement for a circular orbit and in its minimum amount by the requirement for a parabola. From Equation 14.36 these values can be calculated at any particular motion state or velocity condition $v_{location}$. To calculate the required velocity for a parabola the value v_{circle} in Equation 14.36 is replaced by $v_{parabola}$, whereby

$$v^2_{parabola} = 2\, v^2_{circle} \qquad (2.20)$$

Examples of applications for different space missions

- Lunar orbit (Saturn V / Apollo, manned)
- Venus orbit (Magellan)
- Mars orbit (Mars Express)
- Jupiter orbit (Galileo)
- Saturn orbit (Cassini)
- Mercury orbit (BepiColombo)

2.6 Attitude Control and Stabilisation

Every satellite in its orbit requires an attitude and orbit control system (AOCS) for a certain spatial orientation. A fixed position in space is necessary, to ensure a stable alignment of parabolic antennas to the target area. Alignment at a quite specific fictitious position within permissible tolerances, (e. g., within GEO) is also the task of the AOCS. Always existing interferences cause deviations from orbit which lead to areas outside permissible tolerances and have to be actively compensated from time to time.

The most significant interferences of geostationary satellites, which must be deployed in a 'fixed' position above a defined longitude at the equator, are caused by the Moon and the Sun which travel between the tropics at approximately 25° latitude North and 25° South of the equator during the course of the year. They induce a 'shift' of the GEO-satellite from the equator to higher latitudes. The power requirement for station-keeping of these satellites above the equator is designated to as north-south station-keeping (NSSK) and is approximately $150\,\mathrm{m \cdot s^{-1}}$ per year.

Different procedures are used for attitude and orbit control of satellites. Today, the power requirement for such corrective manoeuvres are covered mainly by small chemical propelled rocket engines, but also electrical systems are currently under development. They will in future supplement the chemical ones and eventually will replace them, perhaps one day. In satellite technology, analogous to aviation, rotations around the three axis are also designated to as pitch, yaw and roll. In contrast to aviation, these terms are not standardised in satellite technology because they are defined differently by the producers of propulsion systems. A classification of attitude determination and control systems is listed in Figure 2.50.

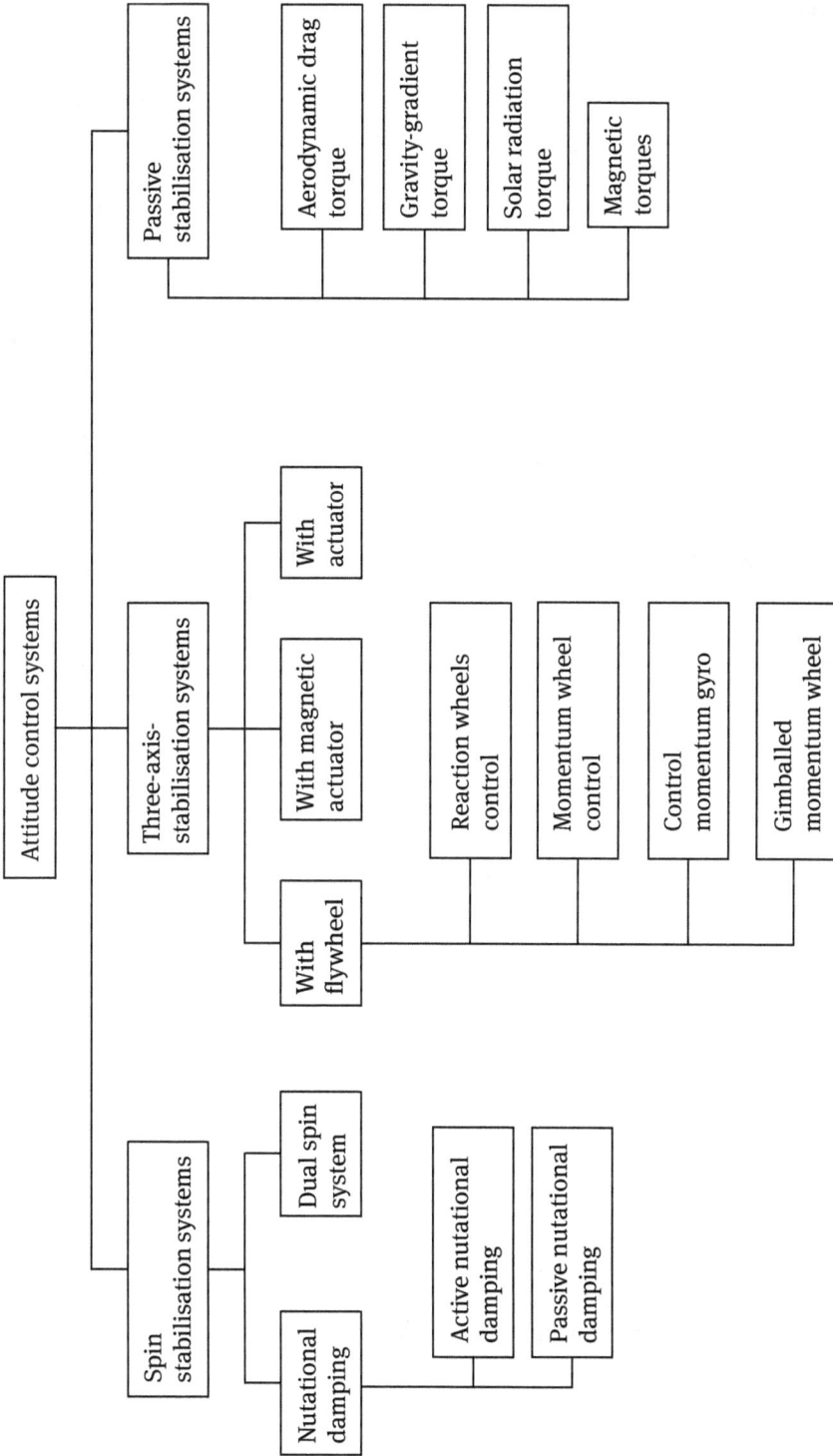

Figure 2.50: Classification of attitude control systems.

2.6.1 Three-Axis Stabilisation

Today the most common way of an active satellite stabilisation is three-axis stabilisation. It is by far the most complex one. There are two forms of gyroscopic three-axis stabilisation. Momemtum flywheels which spin in one direction only or reaction flywheels which spin in both directions.

The realisation of the three-axis stabilisation with the help of fast-running gyroscopes is shown in Figure 2.51. According to the Law of Conservation of Angular Momentum the gyroscopes are striving to maintain their rotary movement and their rotary axis. To some extent, the satellite will therefore be able 'to hold on' to these masses. The gyroscopes get slower over time by internal friction and have to be regularly 'recharged'. To avoid a counter movement of the satellite the orientation of the satellite must be actively ensured by the propulsion system during this time.

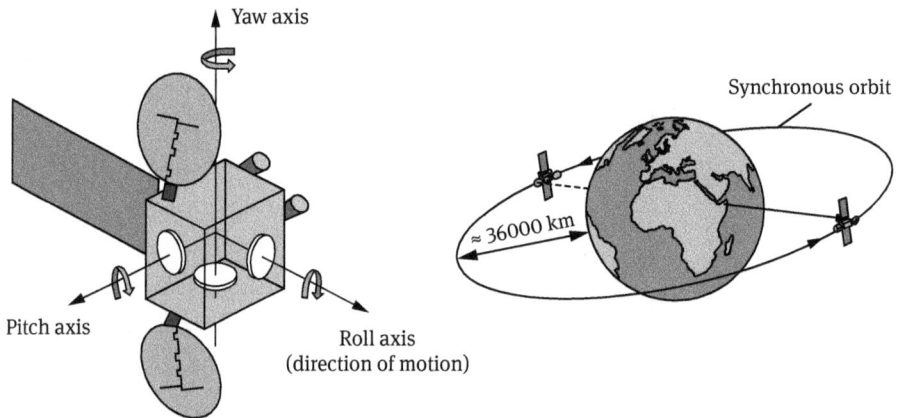

Figure 2.51: Three-axis stabilisation (system with reaction gyroscopes).

2.6.2 Spin Stabilisation

The less extensive spin stabilisation was applied earlier (Figure 2.52). In particular, the whole satellite is set into a defined rotation. To ensure unrestricted accessibility of the satellite and availability of its instruments several parts of the satellite require a reduction of the spin for stationary alignment to a particular target.

The reduction of the spin is quite complicated in such a way that today this stabilisation procedure is seldom used. Please note that the effect of weightlessness is also suspended by this rotation, e. g., because the propellant in the tanks rushes outwards by the resulting centrifugal force. The rotation is also affected by the gravitational forces of the Moon and the Sun which leads to precession and nutation of the rotation axis.

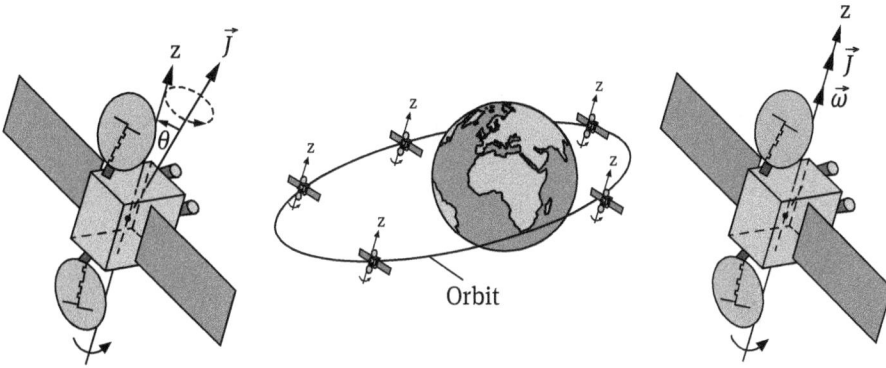

Figure 2.52: Spin stabilisation.

2.6.3 Gravitational Gradient Stabilisation

For larger satellites, structures, and artificial celestial bodies a very simple and reliable stabilisation is passively achieved by displacement of masses. The gravitational force of the central body decreases with the square of the distance. Parts of the satellite which are located closer to the central body are attracted more strongly than others further away. A small gravitational gradient is already sufficient to align rigidly connected bodies (as tethered satellite) or losely connected bodies (as tethered system) in a stable position. For gravitational stabilisation the knowledge of the positions of the following three distinctive points are necessary.

- Centre of mass (CM)
- Centre of gravity (CG)
- Metacentre (MC)

The situation during the separation of a satellite into two parts connected by a tether is shown in Figure 2.53.

$$\omega_{\text{system}} = f\left(\frac{m_1}{m_2}, L_{\text{tether}}, r_{\text{CM}}\right) \tag{2.21}$$

If the case should be

$$\omega_{\text{system}} = \omega_a \tag{2.22}$$

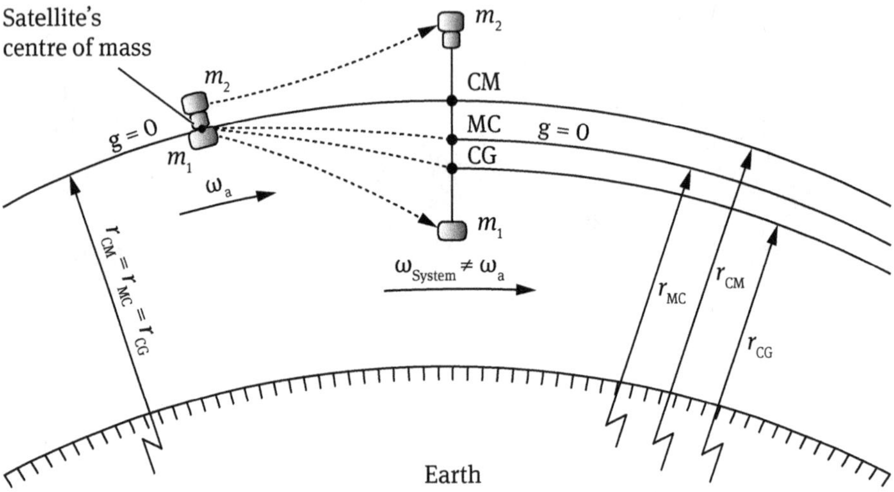

Figure 2.53: Separation of a satellite with tethered connection.

thrust have to be applied onto the system. For a tethered system the three characteristic points mentioned above can be calculated according to the Equations 14.55 to 14.58. There is weightlessness only in the metacentre in terms of figures. All the other points have a residual weight because along the tether an additional force S acts on the partial masses with an acceleration according to Equation 14.60. Also the position of the metacentre is the determining factor for orbit determination and period of orbit (Equation 14.54) of the tethered system.

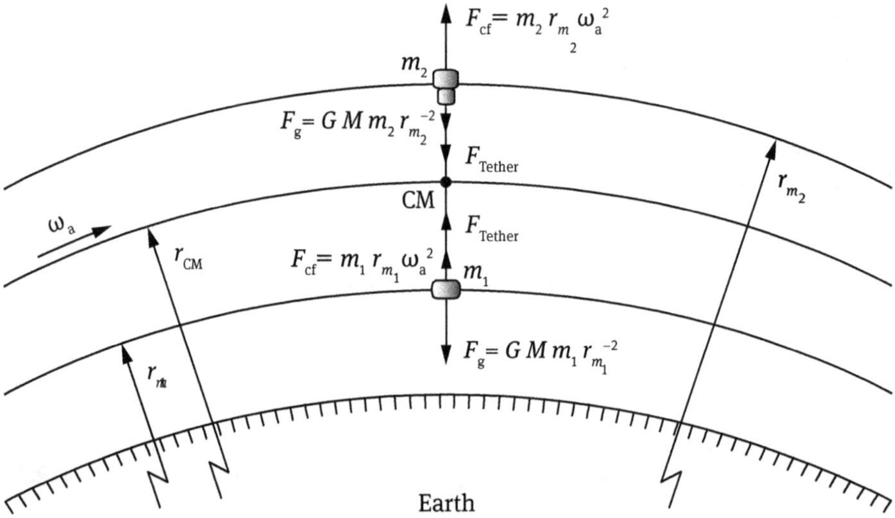

Figure 2.54: Characteristic points CM, CG, and MC for tethered systems.

In contrast to the relatively small satellites accepted as point masses, tethered systems are linear and can reach an enormous size. This may give rise to complications during orbiting of several tether satellites on different orbits, but also due to the complicated mechanisms during unfolding and rolling out of tethered systems. This special space application did not get beyond the experimental stage so far.

2.6.4 Magnetic Stabilisation

Another possibility of passive stabilisation can be applied by using the magnetic field of suitable celestial bodies (Figure 2.55). This type of stabilisation has been applied only for certain scientific applications thus far. The German Azur research satellite rotated twice around its own axis during one orbit around the Earth by orientation along the magnetic lines of force.

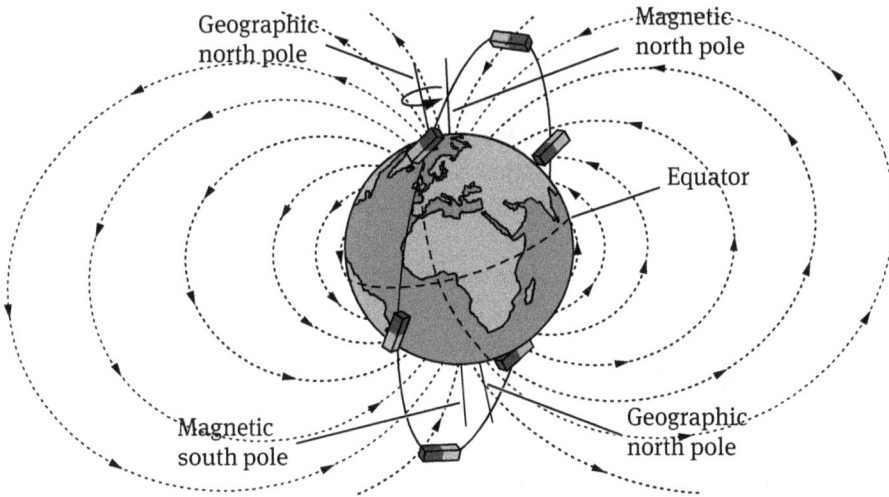

Figure 2.55: Stabilisation by orientation along the magnetic lines of force.

For this type of stabilisation relatively large permanent magnets have to be carried along in the satellite. The intensity of the magnetic field decreases quadratically with increasing distance to the central body. Therefore, the effectiveness of the procedure rapidly meets its limits, Thus, this form of stabilisation is not in use anymore. The magnetic field of the Earth can also be used for the deactivating of reaction wheels in the near-earth orbit. This special method of using the magnetic field of the Earth is applied to the Hubble space telescope.

2.7 Questions for Further Studies

1. Give the names of the nine big planets rotating around the Sun, sorted in ascending order by the length of the semi-axis of their orbits.
2. Calculate the mean density of the Sun, the Earth, and the planet Saturn. Explain the main differences.
3. How long the light is travelling from the Sun to the Earth? How long does it take to receive a confirmation (light running time) of executing control commands from the Galileo Jupiter probe?
4. How long a space probe is roughly travelling from Jupiter to the brightest star, Sirius (at a distance of 8 light years)? The typical velocity of the Space Shuttle is to be assumed as constant cruising speed!
5. Explain why does planet Mercury show phases, whilst planet Jupiter does not.
6. What are the Kepler's Laws of Planetary Motion?
7. What kind of geometric figures can be obtained by intersection of a plane with a circular cone (conic sections)?
8. What force attract two lorries to each other (mass: 40 t each), if their centres of mass are at a distance of 3 m?
9. What planet is more interfering Moon's orbit? Venus or Jupiter, and why?
10. What is an inferior planet and what is a superior planet?
11. What is the weight of a person with a mass of 80 kg on the Sun, on the Earth, and on the Moon?
12. How long a New Year's Eve rocket travels unpowered after burn-out with a burn-out velocity of $30 \, \text{m} \cdot \text{s}^{-1}$ and a launch inclination angle of $45°$ relative to Earth's surface? Duration of propulsion and air drag can be ignored!
13. Why Russian military intermediate-range missiles stationed throughout Europe are bigger and heavier than American or French missiles having the same range?
14. What properties and preferences have sun-synchronous orbits around Earth?
15. What are the characteristics of geostationary orbits?
16. What methods of orbital manoeuvres of satellites do you know? Give advantages and drawbacks.
17. Why our Moon is attracted stronger from the Sun than from Earth?
18. How strong our Sun is attracted from a star of the same mass located at the edge of the universe? (Mass of the Sun: 2E+30 kg, distance from the Sun: 10E+9 light years).
19. Explain the term cosmic velocity. What kind of velocities do you know?
20. What is the fictitious point, the Sun, the Moon and the planets of our Solar System rotate around? Where is the approximate position of this point?

3 Propulsion Systems

All propulsion systems used for spacecraft applications follow the Law of Conservation of Momentum.

$$actio = reactio$$

That means, that each satellite with an attitude and orbit control system and each launch vehicle use fuel which, by repulsion with the highest possible velocity in a certain direction, results in a force, and therefore, in an acceleration in precisely the opposite direction by means of accordingly aligned engines. The propulsion power or the dimensions of a propulsion system can be calculated by using the rocket Equation 14.61 in a first approximation. An exception to this is solar sailing, by which the propellant mass is not carried in the satellite, but supplied from the outside in the form of light particles from the Sun. By absorption and reflection at a mirror face the beam direction of the particles is changed. Some exhaust and supply velocities of different propulsion systems can be seen in Table 3.1.

Table 3.1: Exhaust velocities and supply velocities.

Propulsion	T/K	$v_a/m \cdot s^{-1}$	m_a/kg	tr/kg	$v_e/m \cdot s^{-1}$	Sum
Cold gas (nitrogen)	300	700	100	10	74	
Cold gas (helium)	300	1850	100	10	195	
Single-component (N_2H_4)	900	2100	500	200	1073	
Powder (Al, NH_3ClO_4, ...)	2000	2600	780000	38000	1736	
Storable (MMH, N_2H_4)	3000	3100	15000	9700	3225	
Cryogenic (LH2, LOX)	3200	4300	24000	210000	8942	
Thermal (hydrogen)	2000	7000	10000	8000	11266	
Electric (Xe, ...)	300	30000	10000	8000	48283	
One stage cryogenic (LH2, LOX)	3200	4500	111110	900000	7473	7473
Two stages cryogenic (LH2, LOX)	3200	4500	111100	90000	7475	14949
Three stages cryogenic (LH2, LOX)	3200	4500	111000	9000	7493	22441
Four stages cryogenic (LH2, LOX)	3200	4500	1100	900	7671	30113
Payload of cryogenic propulsion			100			

Where

v_a is the effective exhaust velocity
m_a is the total mass of the propulsion system (example)
tr is the amount of fuel mass (example)
v_e is the supply velocity

Many realised and potential propulsion systems may be roughly divided into the following three groups.

- Air-breathing propulsion systems
- Chemical propulsion systems
- Physical propulsion systems

3.1 Rocket Equation

The derivation of the rocket equation was the first important theoretical step for the technical implementation of space flight. The rocket equation shall be derived from the Law of Conservation of Momentum. Thus,

$$m_1 \cdot v_1 = m_2 \cdot v_2 \tag{3.1}$$

As the rocket may be built up of up to 90 % propellant, and therefore m_1 and m_2 change significantly during firing time of the rocket, the universal Equation 14.61 shall be derived by integration. The mass ratio m_0 / m_e can be described as

$$\frac{m_0}{m_e} = \frac{nl + st + tr}{nl + st} \tag{3.2}$$

Where
m_0 is the mass at the beginning of the firing time
m_e is the mass at the end of the firing time
nl is the mass of the payload
st is the structural mass
tr is the mass of the propellant

For small mass ratios m_0 / m_e the Equation 14.61 can be approximated again to

$$v_e m_e = v_a tr \tag{3.3}$$

This relation is valid for attitude and orbit control of satellites with sufficient accuracy. Hence one can conclude directly on the pursuit of highest possible exhaust velocities of the propulsion systems in order to achieve a propellant saving operation of the attitude and orbit control.

To calculate the maximum transportable payload for an existing propulsion system, the Equations 14.62 to 14.65 shall be derived directly from Equation 14.61 for different missions. However, the prerequisite for this is the re-ignitibility of the propulsion system. With such a space hauler payloads can be transported back and forth between an original orbit and a destination orbit by

- a single flight (without return of the space hauler),
- return of the space hauler without payload to its original orbit,
- an outward flight of the space hauler w/o payload (e.g., pickup of payload), or
- a flight with payload (e. g. with payload for repair to the destination orbit).

3.2 Air-Breathing Propulsion Systems

Air-breathing propulsion systems are used by default for all aircrafts which are heavier than air. Basically, these chemical propulsions use the existing atmospheric oxygen as oxidiser. A distinction is made between propeller drives (with piston displacement engines) and turboprop drives (with gas turbines) both characterised by the performance of the driveshaft. Gas turbines (with higher power characterised by their thrust) are the most frequently used gas turbines in the civil and military sectors. Aircrafts carrying only propellant, such as hydrocarbons, liquefied gas (LNG) in future, and liquid hydrogen (LH2), why these engines in principle are not able to run in space in the absence of oxidisers. Therefore, they are no propulsion systems for outer space. Since all space missions take off from Earth's surface, it makes sense to use air-breathing engines in rocket lower stages and starting aggregates. Today, nearly all launch vehicles take off vertically. Therefore, the aerodynamic drag on the rocket and on the nozzle outlet of the engines causes losses during lift-off. Moreover, during lift-off the thrust must be greater than the fixed weight of the rocket to achieve lift-off. With the help of air-breathing engines and a horizontal take-off by the action of lifting forces of the air there are sufficient thrusts for these propulsion stages which are significantly smaller than of VTOL aircraft.

There are also aircrafts which can take off vertically (e. g. helicopters and VTOL aircrafts in areas where sufficient space is not provided for an adequate take-off and landing strip). However, an aerodynamically supported horizontal take-off is energetically more favourable. The Pegasus air launch system is the only horizontal launcher used in space flight until now. The system is commercially available from Boeing Company for transport of small payloads up to approximately 100 kg. The two-stage Pegasus launcher is mounted as a load under the fuselage of a conventionally modified subsonic carrier aircraft and transported to an altitude of 10 km. After reaching the launch corridor and releasing from the carrier aircraft, the launcher has already a velocity of approximately $300 \, \text{m} \cdot \text{s}^{-1}$ plus Earth's rotational velocity. This system is independent from stationary launch bases. Depending on the orbit, the appropriate starting point can be chosen by the aircraft.

The future Sänger project also uses air-breathing engines in the first stage with horizontal take-off at an appropriate airport or Earth's latitude and subsequent acceleration up to six to seven times the speed of sound to an altitude of approximately 35 km with subsequent stage separation. The upper stage is to be equipped with high-energy cryogenic rocket engines. The Sänger launcher is a completely re-usable payload launch system since both stages fly back to the launch site after finishing the mission. They are then available again for a new mission. For cost efficiency reasons and due to technical problems with hypersonic engines this ambitious project was never realised.

3.3 Chemical Propulsion Systems

The kinetic energy of the recoil mass of chemical propulsion systems originates from the reaction of chemical propellants. Depending on the number of propellants involved a distinction is made between monopropellant propulsions, bipropellant propulsions, hybrid propulsions, and tribrid propulsions. By definition, solid propellant rocket engines are referred to as single-component systems, because the propellant mixture has a homogeneous composition.

Assuming an ideal gas, the exhaust velocity of a gas stream is calculated by the de-Saint-Venant-Zeuner-Wantzel's equation

$$v_a^2 = \frac{2\,\kappa}{\kappa-1}\frac{R \cdot T}{M}\left\{1 - \left(\frac{p_e}{p_0}\right)^{\frac{\kappa-1}{\kappa}}\right\}$$ (3.4)

Where

v_a is the exhaust velocity at the nozzle exit M is the average gas molecular weight

κ is the adiabatic coefficient p_e is the absolute pressure (nozzle exit)

R is the universal gas constant p_0 is the ambient pressure

T is the combustion zone temperature

The adiabatic coefficient κ is a specific property of gases and is approximately $(f+2)/f$, with f as thermal degree of freedom, therefore

– for monatomic noble gases $5/3 = 1.67$,
– for diatomic gases $7/5 = 1.4$, and
– for polyatomic gases $8/6 = 1.33$.

The ratio p_e/p_0 is also referred to as expansion ratio. For lift-off from ground the pressure at the exit of the nozzle shall not be chosen too small to avoid disturbing flow separation at the exit of the nozzle. Typically, the pressure at the nozzle's exit is greater than 0.3 bar. A combustion chamber pressure of 100 bar results in an expansion ratio of 300!

The ratio of the area at the nozzle exit to the narrowest cross-section is referred to as expansion ratio. With increasing expansion ratio the specific impulse increases. However, there are limits for this value by structural restrictions. The thrust of a chemical propulsion system is calculated from

$$F = d_m\, v_a + \left(p_e - p_0\right) A$$ (3.5)

Where

F is the thrust
d_m is the flow rate (fuel consumption)
A is the area of the nozzle exit

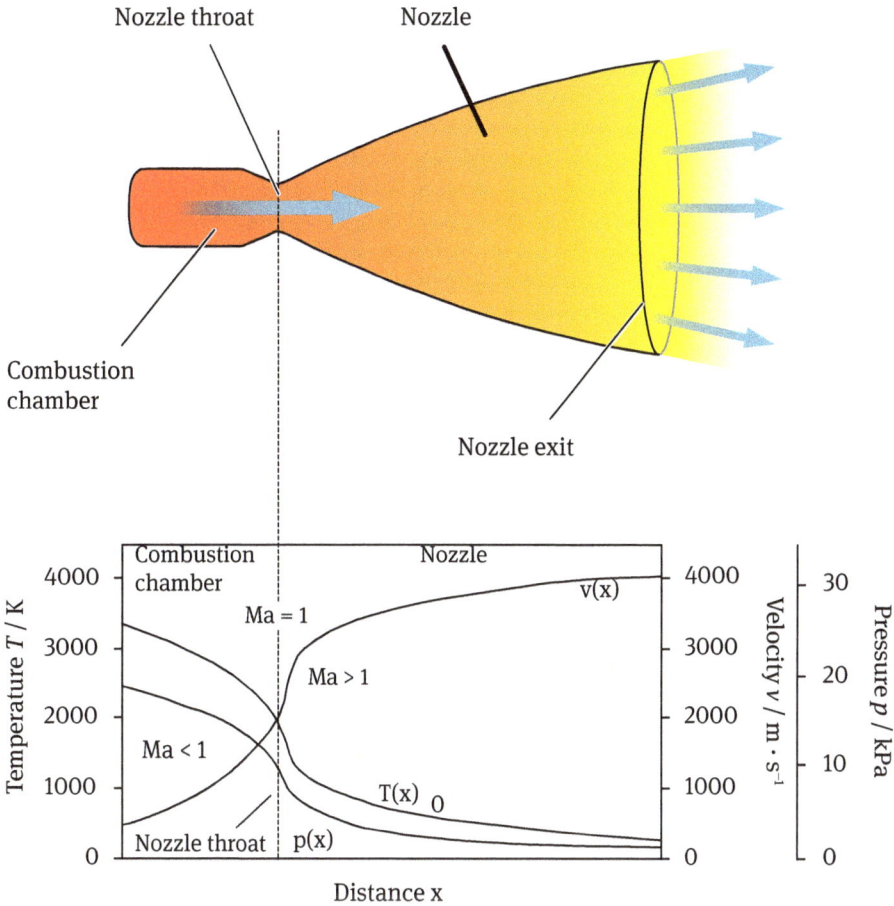

Figure 3.1: Data from flow simulation of a de Laval nozzle.

Because of the pressure term $p_0 \cdot A$ at the end of the Equation 3.5, the thrust of the engines in a vacuum is always greater than during lift-off from ground. The following Figure 3.1 shows the typical course of flow data within a de Laval nozzle. The shape of a de Laval nozzle depends from the gas-dynamic properties of flowing media. After injection and combustion of propellants in the subsonic range (Mach number Ma < 1) an acceleration of the flow is achieved if the flow cross-section narrowed (converges) while sound velocity Ma = 1 is adjusted in the narrowest cross-section. A further acceleration into supersonic range (Ma > 1) is achieved if the cross-section is expanded in the flow direction (diverges).

Rocket propellants with very high combustion energy and a low molecular weight of the combustion products are suitable for chemical propulsions (Table 3.2). A selection of propellant combinations is shown in Table 3.3.

Table 3.2 Reaction products with high heat of formation (s is the solid, g is the gaseous).

Chemical formula	Heat of formation / kJ·kg^{-1}	F_P/°C	K_P/°C
BeO	23 950 (s)	+2550	+3850
LiF	23 590 (s)	+848	+1767
B$_2$O$_3$	18 340 (s)	+450	+2217
BF$_3$	16 660 (g)	– 129	– 101
Al$_2$O$_3$	16 410 (s)	+2045	+2700
MgO	14 950 (s)	+2642	+2800
HF	14 180 (g)	– 85	+19
H$_2$O	13 430 (g)	0	+100
BN	9 800 (s)		(+2327)
CO$_2$	8 960 (g)	– 58	– 79 (subl.)

Table 3.3: Selection of propellant combinations.

Fuel	Oxidiser	Application
Ethanol (CH$_3$-CH$_2$-OH)	Liquid oxygen (LOX)	Aggregat-4 (V2)
Kerosene (RP-1)	Liquid oxygen (LOX)	1st stage of the Saturn V rocket
UDMH (CH$_3$-N$_2$H$_2$-CH$_2$)	Dinitrogen tetroxide (N$_2$O$_4$)	1st and 2nd stage of the Ariane 4 rocket
MMH (CH$_3$-N$_2$H$_3$)	Dinitrogen tetroxide (N$_2$O$_4$)	Upper stage of the Ariane 5 rocket (expired)
Liquid hydrogen (LH2)	Liquid oxygen (LOX)	– 2nd and 3rd stage of the Saturn V
		– Energiya
		– 3rd stage of the Ariane 4 rocket
		– Main stage of the Ariane 5 rocket
		– Upper stage of the Ariane 5 rocket (in future)
		– Atlas 5 rocket
		– Delta 4 rocket

The quality of a chemical propulsion system is usually expressed by the parameter of the specific impulse. Thus, it is

$$I_{sp} = \frac{v_a}{g} \tag{3.6}$$

Where

I_{sp} is the specific impulse
v_a is the exhaust velocity
g is the Earth's acceleration (9.81 m·s^{-2})

The unit of the specific impulse I_{sp} is of time and thereby, the following definition is applied. The specific impulse I_{sp} represents the firing time of a reference engine generating a thrust of 1N by consuming 1kg of propellant. Sometimes the specific impulse is inaccurately referred to as exhaust velocity of the propellants c_e. The difference is then recognisable only by the unit.

The power of chemical rocket propulsions is a rather untypical characteristic. The propulsive power P of an engine is calculated as the product of thrust (force) F and exhaust velocity v_a.

$$P = F \cdot v_a \qquad (3.7)$$

Each of the five engines of the first stage of the Saturn V rocket achieved a propulsive power of $21 \cdot 10^9\,\text{W} = 21\,000\,\text{MW}$ or 28 million horsepower at a thrust of 700 t and an exhaust velocity of approximately $3000\,\text{m}\cdot\text{s}^{-1}$. The engines of the Saturn V lunar rocket were the most powerful engines ever developed and used. The following Figure 3.2 shows the principle design of chemical propulsion systems, divided by the stage of aggregation of the propellants.

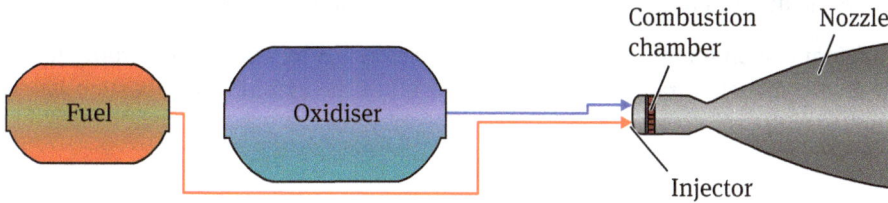

Liquid Propulsion System

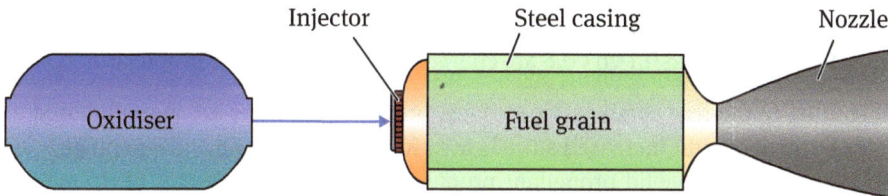

Hybrid Propulsion System

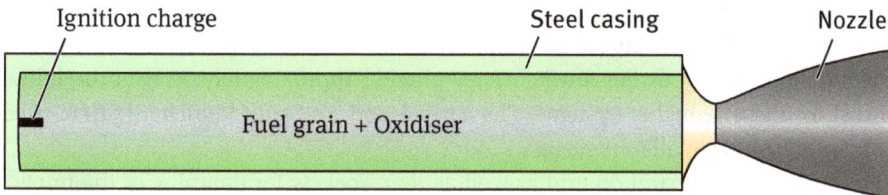

Solid Propulsion System

Figure 3.2: Schematics of chemical propulsion systems.

3.3.1 Solid Propulsion Systems

Many historians consider the invention of gunpowder in late-medieval China as the beginning of space flight. There are reliable records on a Chinese Mandarin being the first taikonaut (term for Chinese astronauts) who died early on by accident. Today, black powder contains approximately 75 % potassium nitrate, 15 % charcoal, and 10 % sulphur. Still existing in pyrotechnics as propulsion unit, it is often used in combination with metallic elements for colourful fireworks. However, the exhaust velocity in rocket engines is not particularly high. Modern-day solid propulsion systems are typically composed as follows.

- NH_4ClO_4 or NH_4NO_3 as oxidiser (over 70 %)
- Plastics (polybutadiene, polyurethane, polyacrylonitrile) as fuel and binder (approximately 15 %)
- Admixtures of light metals such as Al, Mg, Li, and Be (up to 15 %)

Different combinations can be prepared (typically 0.5 to 5 cm·s^{-1}) according to the desired combustion rate. For a combustion rate r the empirical combustion law is applied with

$$r = a \cdot p^n \tag{3.8}$$

Where

r is the combustion rate
p is the combustion chamber pressure in bar
0.2 < a < 8 mm·s^{-1} and 0.1 < n < 0.8 are characteristic figures of the propellant.

According to the free combustible surface, propelling charges are manufactured as front burners or with increased surface area in the form of conical deepenings (core of fireworks). Large solid propellant propulsion systems with length of up to 50 m are used with cylindrical surface. They have a circular or radial cross-section. Solid propellant propulsion systems are widespread in space flight for their simple design with only a few moving parts. They are used in very different applications. The largest solid propellant rocket engine serves as launching aid for the American Space Shuttle and the European Ariane 5 launch vehicle. Former engineers called them quite true as 'powder towers'. They provide approximately 90 % of the initial thrust for these launcher systems. The early American Scout launcher system used solid propulsions in all four stages.

During stage separation and as well as retro-rockets for damping of the impact of the touch-down of Russian Soyuz capsules during return from space, these propulsion systems are also running highly reliable. The top of the Saturn V rocket also carried a booster which served as a rescue facility in case of emergency for the three astronauts in the Apollo capsule. Fortunately, it was never used therefore.

The exhaust velocities of modern solid propellants are of the order of $2400\,\text{m}\cdot\text{s}^{-1}$ to $2900\,\text{m}\cdot\text{s}^{-1}$. The application possibilities are limited to one-time use because re-ignitability is not possible. During launch of the Space Shuttle or Ariane 5 rocket the boosters are separated from the main stage typically two minutes after launch. Floating on parachutes they fall down into the waters of the Atlantic Ocean at a distance of approximately 1000 km to 2000 km away from the launch site. The boosters are in principle suited for reuse, however, there are no cost benefits over disposable boosters because of comprehensive inspections. Due to a remaining mass of 80 t per booster of the Space Shuttle the largest currently available parachutes are required for splashdown. The main advantages of these propulsion systems are their almost unlimited storability and their prompt availability which is required for military purposes.

3.3.2 Liquid Propulsion Systems

In 1926, *Robert H. Goddard* already started the construction of the first liquid propulsions. Today, liquid propulsion systems are the most commonly used propulsion systems. The supply of propellant is necessarily achieved by pumps at high pressures in the combustion chamber, while in the low pressure range also more reliable and more cost-effective compressed gas supply is used. According to the number of components a distinction is made between monopropellant and bipropellant propulsion systems.

There are liquid propulsion systems which are re-ignitable and other systems which do not have that capability. Propulsion systems which are re-ignitable can be operated by steady state firing (SSF) or by pulse mode firing (PMF). Very small thrusts are required to some extent for attitude control of satellites which are achieved by PMF operation. These propulsion systems require fast-switching valves in the range of 10 ms.

Monopropellant Propulsion Systems

The relative instability of suitable chemical compounds is used by pure mono-propellant propulsion systems. The use of hydrogen peroxide (H_2O_2) was of importance in the past. Today, hydrazine (N_2H_4) is used almost exclusively in monopropellant propulsion systems with a typically exhaust velocity of up to $2\,100\,\text{m}\cdot\text{s}^{-1}$ because hydrogen peroxide has a sensitivity to shocks. Today, these systems are also used to a greater extend as emergency surfacing systems in submarines because a large amount of gas is generated within a short period of time. Water is pressed out of the ballast tanks, thus resulting in rapid surfacing.

Monopropellant systems are more reliable and cheaper than bipropellant systems since less units of all components are required (tanks, valves, pumps, etc.), and an ignition device can be omitted. But the specific impulse is worse than with bipropellant systems. Today, these systems still play a special role for attitude and orbit control of small satellites.

With Ariane 5 EPS upper stage these thrusters are also used for attitude control (SCA) after separation of the solid propellant boosters. The operation is based on the catalytic decomposition of hydrazine. Special platinum metals (iridium, ruthenium) are used as catalysts which are applied onto a carrier material (e. g. Al_2O_3) with a large effective surface. Ammonia (NH_3), nitrogen (N_2), and hydrogen (H_2) are released as combustion gases, whose composition, and therefore the specific impulse I_{sp}, depends on pressure and temperature.

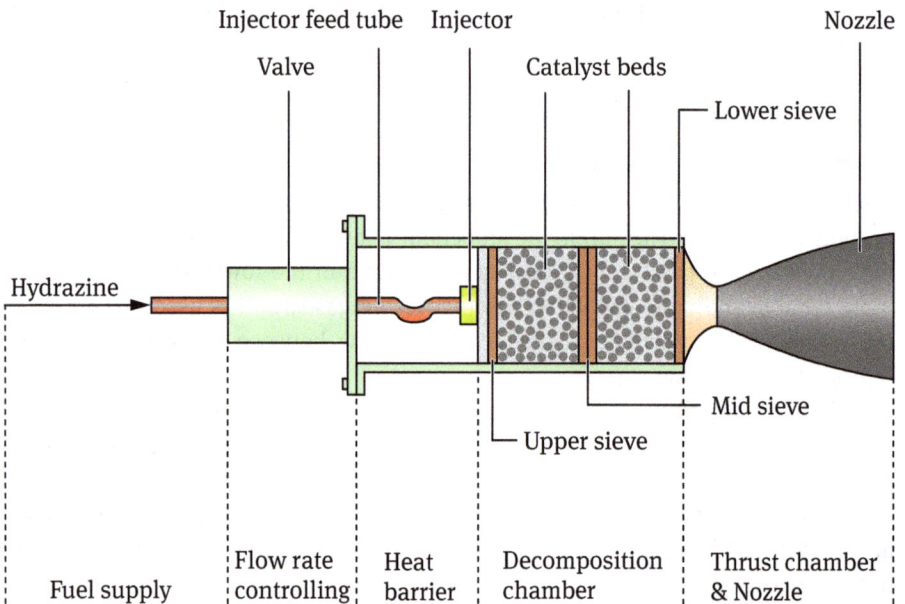

Figure 3.3: Principle design of a monopropellant thruster.

Monopropellant thrusters are used for thrusts from 0.5 N up into the 100 N range. With a thrust of 1 N a nozzle throat diameter results in less than 1 mm, which are the smallest chemically propelled rocket engines used.

Bipropellant Propulsion Systems

Bipropellant propulsions are the most frequently used propulsion systems in space flight. Hydrogen/oxygen propulsion systems were used early due to their high energy density. Cryogenic propellant components are filled in tanks in liquid form shortly before lift-off. After launch the propellant components are pumped or alternatively pressed to the engines (helium into hydrogen tank and gaseous oxygen into oxygen tank).

Cryogenic LH2/LOX-propulsion systems have exhaust velocities up to $4\,600\,\text{m}\cdot\text{s}^{-1}$. They are used for the American Space Shuttle and for the main stage of the European Ariane 5 rocket. The engines of the Space Shuttle are re-ignitable (electric ignition), while the HM60 engine of the Ariane 5 rocket can be ignited only once (pyrotechnic ignition).

Also hypergolic mixtures of fuel are of great importance. Hydrazine derivatives (MMH, UDMH or hydrazine mixtures) are used as fuel, and dinitrogen tetroxide (N_2O_4) is used as oxidiser. Both components can be stored for an almost unlimited period of time and need no ignition device. Since hypergols are liquid at normal temperatures, they are easier to handle and store than cryogenic propellants. However, hypergols are highly toxic and must be handled with extreme care. Hypergolic propellants are ideal for spacecraft manoeuvering systems due to their easy start and restart capabilities.

The hypergolic reaction starts immediately upon contact of the components. Typically, hypergolic propulsion systems have exhaust velocities up to $3\,200\,\text{m}\cdot\text{s}^{-1}$. They are referred to as medium energetic propulsions. Its scope of application ranges from 10 N (for satellite attitude control) up to several hundred tons of thrust in lower stages of Russian launchers and in the former Ariane 4 rocket.

Table 3.4: Properties of selected fuels for liquid propellants.

Fuel	Melting point / °C	Boiling point / °C	Density / $g\cdot cm^{-3}$	Comment
Hydrogen	− 259.1	− 252.7	0.08	High power, low density, not storable
Hydrazine	+ 1.7	+ 113.7	1.01	Toxic, explosive
UDMH	− 58.0	+ 63	0.81	Toxic
MMH	− 52.4	+ 87.5	0.876	Toxic
Kerosene	− 4.3	≈ 200	0.8	RP-1
Ethanol	− 114.1	+ 78.3	0,78	–

Table 3.5: Properties of selected oxidisers for liquid propellants.

Oxidiser	Melting point / °C	Boiling point / °C	Density / $g\cdot cm^{-3}$	Comment
Oxygen	− 218.7	− 183	1.14	Not storable
N_2O_4	− 11.2	+ 21.2	1.45	Toxic
Nitric acid	− 51.6	+ 84.0	1.52	Toxic
Fluorine	− 219.8	− 188.1	1.51	Extremely corrosive
Ozone	− 184	− 110	1.57	Sensitive to shock
H_2O_4	− 0.43	+ 150	1.45	Sensitive to shock

Depending on the method of combustion, a distinction is made between closed cycle expander engines and open cycle expander engines. In closed cycle expander engines all fuels are supplied to the combustion chamber, thus contributing to thrust generation. In open cycle expander engines part of the fuels is used to drive the turbo units or nozzle cooling.

Closed cycle expander engines have a higher specific impulse than open cycle expander engines, but have higher technical requirements. An operating schematic of bipropellant liquid engines is shown in the next figure. During firing of the engines heat flux densities up to $10\,kW\cdot cm^{-2}$ occur in the region of the narrowest cross-section. Specific measures are required to avoid fracture mechanical failure of the engines. Several cooling methods for engines are available.

– Regenerative cooling
– Film cooling
– Radiative cooling
– Ablative cooling
– Capacitive cooling

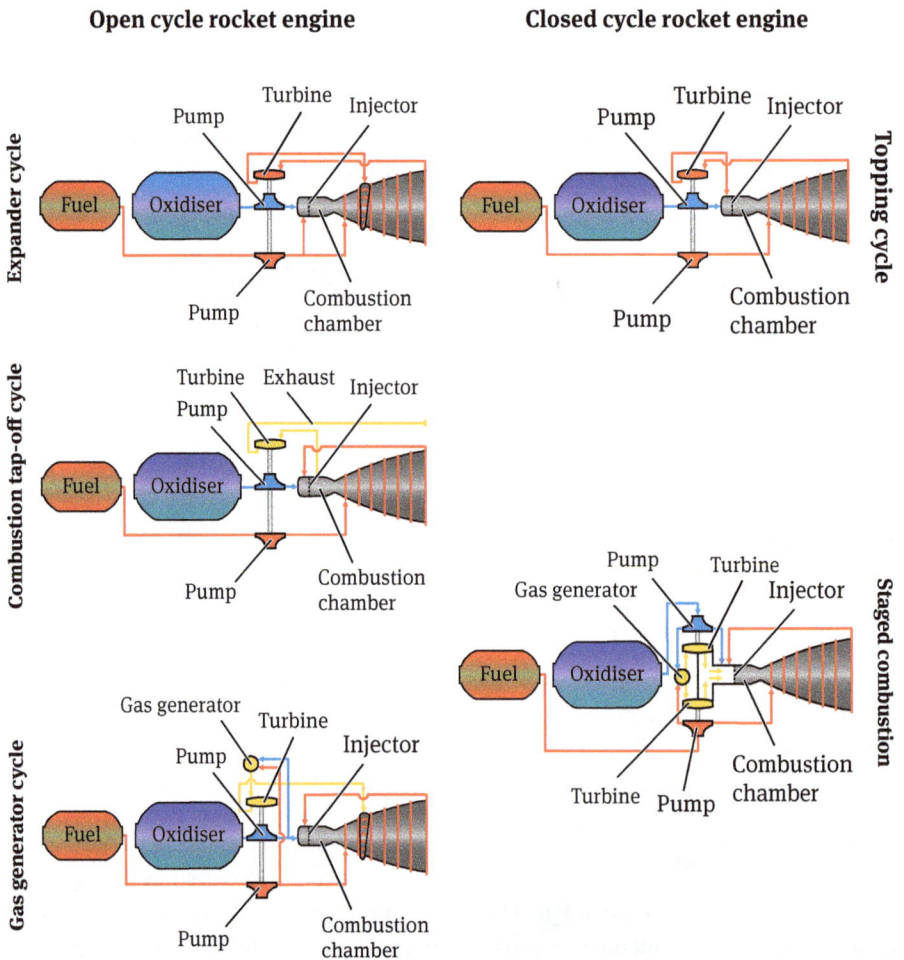

Figure 3.4: Engine systems.

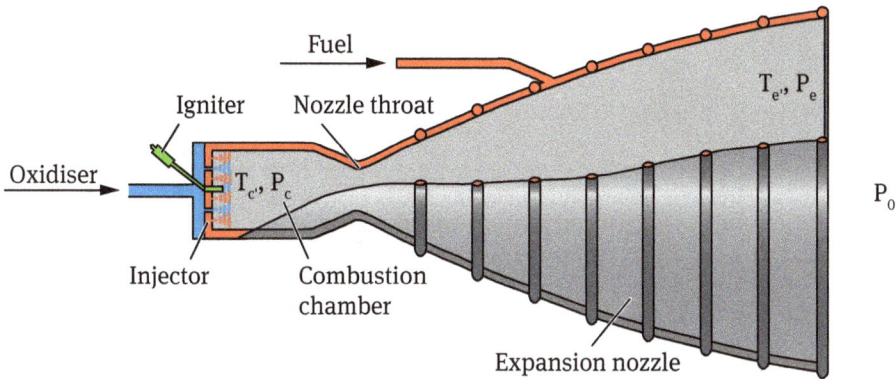

Figure 3.5: Principle setup of a regeneratively cooled engine.

All cooling methods cause a loss of efficiency of the engines, with the exception of regenerative cooling. The thermal and mechanical parameters are important factors of material selection for high-performance propulsion engines. While copper alloys (with zirconium and silver content) are used in large-sized engines with regenerative cooling, platinum alloys (with rhodium, rhenium, or iridium) or Haynes alloys (Co55Cr20W15Ni10) are used in small engines with radiative cooling. The output propulsive power is thereby approximately proportional to the fourth power of the temperature. At a typical temperature of 1000 °C this results in a cooling capacity of

$$P = \sigma \cdot T^4 \tag{3.9}$$

$$P = 5.67 \cdot 10^{-8} \left(1000 + 273\right)^4 = 14.9 \; \frac{W}{cm^2}$$

Alongside the problem of cooling of highly stressed parts in high-performance engines (e. g. the wall of the combustion chamber in the area of the nozzle throat and high-speed turbo pumps for propellant supply), the phenomenon of thermal instabilities plays a major role. Combustion instabilities are a permanent threat during ignition and operation of chemical rocket engines. They have an impairing effect on performance and may have as a consequence that the combustion chamber is destroyed completely by high material wear. This leads to a total loss of the mission. Combustion instabilities mainly occur in high-energy propellants and hypergolic propellants. Possible reasons could be related to locally high energy release in association with acoustic and aerodynamic properties. In practice, a provision is used to prevent or to reduce the instabilities with tuneable Helmholtz resonators (cavities) around the combustion chamber or with baffle plates which extend into the combustion chamber. Despite of the urgency of the problem insufficient progress has been made beyond trial and error.

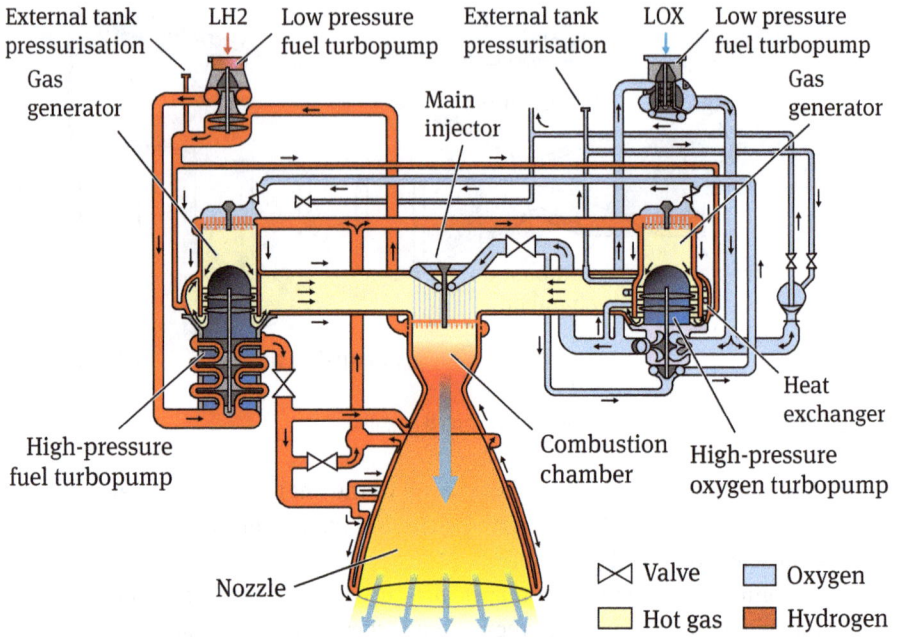

Figure 3.6: Space Shuttle main engine (SSME, LH2/LOX) with staged combustion. Credit: NASA [7]

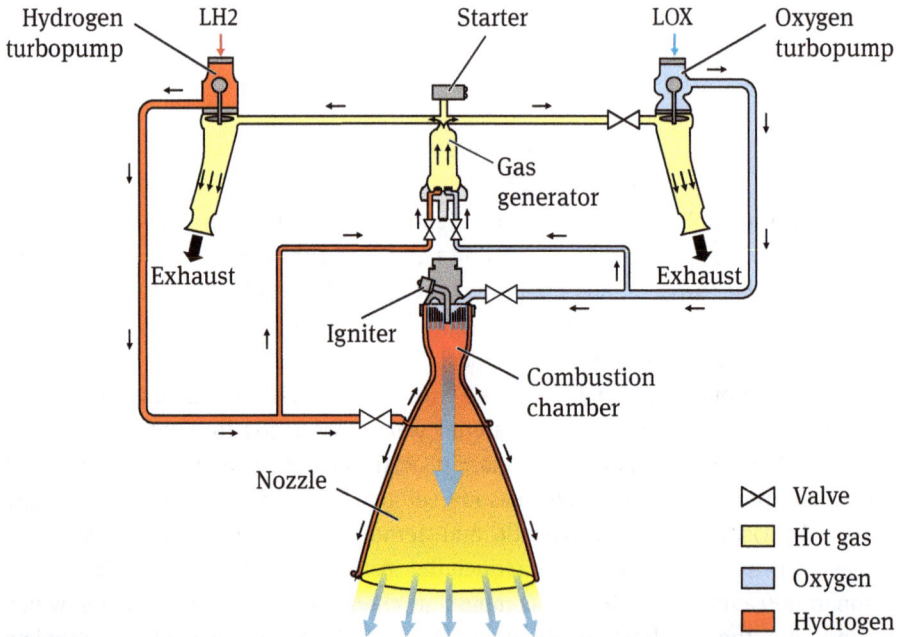

Figure 3.7: Vulcain main engine of Ariane 5 (HM-60, LH2/LOX) with gas generator [8].

Detailed modelling of flow and combustion processes is an important task in science and industry to solve this problem. The current state still does not provide a sufficient basis to save expensive large-scale experiments in future. In addition to high-frequency combustion instabilities there are also low-frequency disturbances (so-called pogo oscillations) during operation of rocket engines. These disturbances are mostly caused by oscillations in the fluid area of the fuel supply system (e. g. swashing of fuel within the tanks). Special pogo suppression devices for suppression of dangerous low-frequency oscillations are used for this.

3.3.3 Hybrid Propulsion Systems

Hybrid thrusters are characterised by the fact that one of the propellant components, mostly the oxidiser, is used in liquid form and the other component is used in solid form. Such propellant systems are also referred as lithergols. Analogous to solid propellant thrusters, hybrid thrusters are composed of a cylindrical container which serves as storage tank and combustion chamber at the same time. Supply of the liquid propellant component is usually achieved by an injection system in the combustion head. A pressure conveying system is mostly used for this as pump installations make the system more complex. Theoretically, the power output may reach the range of highly energetic liquid thrusters. Their extensive safety against explosive combustion processes is of advantage since burn-off of the solid component can only occur to the extent as the liquid component is supplied and a combustible surface of the solid component is available at the same time. Thereby hybrid thrusters are exceptionally easy to control and provide favourable conditions for repeated shutdown and ignition.

3.3.4 Tribrid Propulsion Systems

Tribrid thrusters are supplied with propellants which are composed of three components (so-called tri-ergols). Their structure is similar to conventional liquid-propellant thrusters. If one of the two fuels is in solid form, they are similar to hybrid thrusters. Very highly energetic propellant mixtures (e. g. fluorine-lithium-hydrogen) have very high combustion temperatures up to $4800\,K$ which then have to be lowered by injection of additional hydrogen. Tribrid thrusters have the highest specific impulses of all chemically propelled rocket engines with exhaust velocities of a maximum of $5000\,m \cdot s^{-1}$. Due to high technical requirements and problematic environmental compatibility these propulsion systems are not used in commercial aerospace.

3.4 Physical Propulsion Systems

In the case of physical propulsion systems the kinetic energy of the recoil mass is not generated by the reaction of chemical propellants. Despite of solar sailing as a special case, a distinction according to the energy source is made in

– cold gas propulsion systems,

– electric propulsion systems, and

– thermonuclear propulsion systems.

3.4.1 Cold Gas Propulsion Systems

Figure 3.8: Manned manoeuvring unit (MMU) as independent satellite. Credit: NASA , resized [9].

The most prominent application of a cold gas propulsion system in space technology is the manned manoeuvring unit (MMU). Analogous to chemical propulsions a gas carried along in pressure cylinders (nitrogen or helium) is relaxed and cooled in a de Laval nozzle. The de-Saint-Venant-Zeuner-Wantzel's equation from Chapter 3.2 shall apply *mutatis mutandis*. In principle, a blown-up and let gone balloon is also a simple cold gas propulsion component. Due to limited propulsion power these propulsion systems are only used in small engines with low propulsion requirements (e. g. fine adjustment of the orbit and attitude control of satellites). They are mentioned here for the sake of completeness. The advantages of these systems are simplicity, and therefore, high reliability and low costs.

Nitrogen is used with an exhaust velocity of $v_a = 700 \, \text{m} \cdot \text{s}^{-1}$ if thrust-optimised requirements are favoured, otherwise helium ($v_a = 1800 \, \text{m} \cdot \text{s}^{-1}$) is used for weight-optimised requirements. In the military arena cold gas propulsion systems are used for launching intercontinental missiles from submerged submarines. Carbon dioxide is used as propellant. Ignition of the rocket engines happens after ascent above the surface of the water.

3.4.2 Electric Propulsion Systems

In the case of electric propulsion systems the mass particles of the propellants receive their energy by an external impact through electrical energy. Although they provide exhaust velocities which are roughly above chemical propulsions by a factor of 10 or more, the thrust achieved remains within the sub-Newton range. In principle, the reason is an exceptionally low mass throughput. These engines cannot be used as boosters for launch vehicles, but only for continuous operation in the high vacuum of space. During missions extending over several years exceptional delta velocities are possible. For example, a Mercury probe with a mass of 1000 kg and an average propulsion power of 5 kW with a specific impulse of 5000 s has a maximum propellant consumption according to

$$p = \frac{m}{2t} v_a^2 \tag{3.10}$$

to 15 g per hour and thereby a thrust of

$$F = \frac{m}{t} v_a = 200 \, \text{mN} \tag{3.11}$$

and thereby again an acceleration of

$$a = \frac{F}{m} = 2 \cdot 10^{-4} \, \frac{\text{m}}{\text{s}^2} \tag{3.12}$$

after launch and

$$a = \frac{F}{m} = 3 \cdot 10^{-4} \, \frac{\text{m}}{\text{s}^2} \tag{3.13}$$

at its destination.

For missions from an Earth-like orbit with an inclination of approximately 7° (inclination corresponds to the target orbit around Mercury) into a Mercury-like orbit a rough velocity requirement of

- 16 000 m · s⁻¹ (for Hohmann 2-impulse transfer) and
- 17 000 m · s⁻¹ (for spiral transfer) is required.

A spiral transfer results from these low accelerations with a period of propulsion of over two years. Furthermore, that implies a medium radial approach to the Sun with a velocity of 4 500 km · h⁻¹. This results in several orbits around the Sun during downward spiralling until the orbit of Mercury has been reached. During this period the propellant consumption amounts to 300 kg of Xenon in this sample calculation. At the beginning of the mission another velocity requirement is needed for correct orbit inclination.

$$2\, v_{Earth}\, \sin\left(\frac{7°}{2}\right) = 2 \cdot 30000\, \frac{m}{s} \sin 3.5° = 3660\, \frac{m}{s} \tag{3.14}$$

Also downward spiralling into the gravity field of Mercury requires a velocity of up to 3000 m · s⁻¹. If the two parts of the manoeuvre are performed by electric engines the flight time (with continuous operation of the engines) takes off over three years. A xenon proportion of the propellant of over 50% of the probe's launch mass is required. The enormous supply velocity (by large I_{sp}) is the driving force of further development of electric propulsion systems for long-term missions. The realisation of the calculated Mercury mission is almost impossible with chemical propulsions. Next to specific impulse the power consumption is a characteristic to describe electric propulsion systems. Therefore, they are also referred to as power propulsion systems, in contrast to thrusters where chemical power only plays a subordinate role. Electric propulsion systems can be divided into three groups, and the transitions are smooth to some extent.

- Electrothermal propulsion systems
- MPD or plasma propulsion systems
- Electrostatic propulsion systems

Electrothermal propulsion systems

The propellant of an electrothermal thruster is heated by electrical energy. A cold gas thruster becomes an electrothermal thruster by using an additional heating. A distinction is made between two different principles.

- Resistance heating
- Electric arc heating

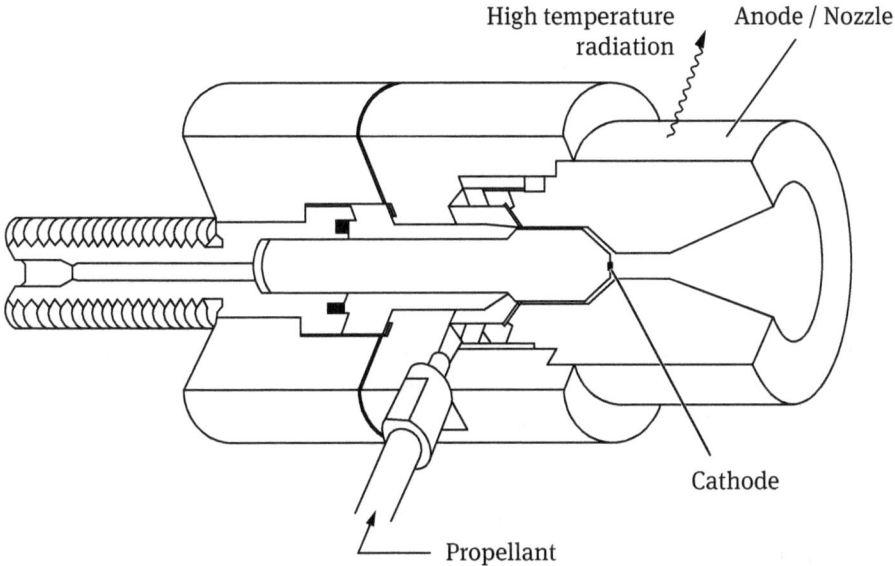

Figure 3.9: Schematic representation of an arcjet.

Since the de-Saint-Venant-Zeuner-Wantzel's equation from Chapter 3.2 is also applied to these thrusters, the propellants should have the smallest possible molecular weight. The principle of resistance-heated thrusters (resistojets) is that resistor bodies are being heated and energy in the form of heat is added to the ambient propellant. Limited by the temperature of the walls, the specific impulse of the thrusters amount to up to $1000\,s$ ($=10\,000\,m \cdot s^{-1}$ exhaust velocity). The thrusters have achieved a high level of development and are used partly for attitude control of satellites.

The basic thermal equations previously mentioned have to be expanded by electromagnetic terms for description of arcjet thrusters because ohmic heating and electromagnetic volume forces must be taken into account. Ohmic heating is primarily used to generate thrust in arcjets, since the amperages in the electric arc are so low that the Lorentz forces do not provide for any significant share of the total thrust. When designing the system care is taken to ensure that the effective anode does not significantly exceed the narrowest cross-section because the ohmic heating has an accelerating impact only in the subsonic range of the flow.

Within an arcjet a propellant is heated by an electric arc. The propellant changes its state from liquid to gas and then goes out of the nozzle. Typical propellants are hydrogen, ammonia or hydrazine. Depending on temperature and propellant an exhaust velocity of $5000\,m \cdot s^{-1}$ to $15000\,m \cdot s^{-1}$ is feasible. The power of an arcjet is in the range of $1\,kW$ to $30\,kW$.

Magnetoplasmadynamic propulsion systems

There is a smooth transition from arcjets to plasma thrusters. The utilisation of electromagnetic forces to increase the specific impulse is the major focus of magnetoplasmadynamic (MPD) thrusters. The ohmic heating plays a subordinate, but not negligible role. A distinction is made between three main types.

- Self-field thrusters
- Applied-field thrusters
- Hall effect thrusters

In self-field thrusters the self-induced magnetic field generated by the current in the plasma is used for acceleration. Typical thrusters, using argon as propellant, achieve a thrust up to 100 N at a power consumption of over 100 kW! In the case of the applied-field thrusters an external axially aligned magnetic field, additional to the self-field, is used to increase the specific impulse. This magnetic field is generated by permanent magnets or magnetic coils, whereas the applied field is significantly stronger than the self-field. Typical thrusters, using preferably light gases as propellant, achieve a thrust up to 1 N at a power consumption of several kilowatts and a specific impulse of 3 000 s.

Hall effect thrusters have an external magnetic field like applied-field thrusters, which is primarily radial aligned. The plasma of these thrusters is generated in different ways, like in electrostatic thrusters as discussed later on (e. g. radio frequency ionisation or glow discharge ionisation).

The charged particles show a drifting motion due to the magnetic fields. Lightweight electrons move almost steady on circular paths while ions hardly deflected in azimuthal direction due to their heavy mass. Thrust is generated by axial acceleration of the ions caused by the electric field applied. The ions have to be neutralised behind the engine to avoid space charges. This type of thruster is thus a transition to electrostatic thrusters. Typical thrusters, using xenon or mercury as propellant, achieve a thrust in the mN range and a specific impulse of 3 000 s at a power consumption up to a maximum of 5 kW. In Russian space flight several Hall effect thrusters are flight proven and used for many years.

Electrostatic propulsion systems

A plasma is generated in electrostatic thrusters by accelerating ions with axial electric fields. The ions leave the thruster and are outside neutralised by electrons to avoid space charges. The individual types of thrusters are different in generating plasma. This can be achieved by

- electron impact ionisation,
- an electric arc (Kaufman thruster), or
- radio-frequency (RIT thruster)

The field emission electric thruster is a form of ion thruster that uses liquid metal (caesium or indium). A high potential difference is applied between two electrodes which extracts ions directly from the liquid metal (electrostatic atomisation).

Method of operation

The major part of the electric output is transformed into kinetic power of the exhausted particles during an approximate loss-free operation of electrostatic thrusters. The rest of approximately 10 % is used for ionisation of the propellant. A small amount in the percent range is used for impacting particles on the accelerating grids which thus do not generate any thrust.

$$P = q \cdot V = \frac{m}{2 v_a^2} \tag{3.15}$$

Where

q is the charge of particles
V is the voltage
m is the mass of particles
v_a is the exhaust velocity

This results in an exhaust velocity of

$$v_a = \left\{ 2 \frac{q}{m} V \right\}^{\frac{1}{2}} \tag{3.16}$$

A throughput of

$$\frac{m}{t} = \frac{m}{q} I \tag{3.17}$$

results in a thrust of

$$F = I \left\{ 2 \frac{q}{m} V \right\}^{\frac{1}{2}} \tag{3.18}$$

Equation 3.18 shows that the thrust is proportional to the ion current I and to the square root of the acceleration voltage V. To achieve a high thrust, heavy ions should be used (mercury or xenon). However, the specific impulse is reduced.

Unfortunately, the ion current I depends not only on the yield of the ioniser, but it is further limited by building up space charges. Particles with the same charge, here ions of the propellant, repel each other as a result of Coulomb interaction. Therefore, the maximum ion current I_{max} can be put through a beam cross section A

with an acceleration voltage V applied and a length d. According to the Child-Langmuir Law this maximum ion current I_{max} is represented by

$$I_{max} = \frac{4}{9}\varepsilon_0 \left\{2\frac{q}{m}\right\}^{\frac{1}{2}} V^{\frac{3}{2}} \frac{A}{d^2} \tag{3.19}$$

Where

ε_0 is the dielectric constant $(8.85 \cdot 10^{-12}\,F \cdot m^{-1})$

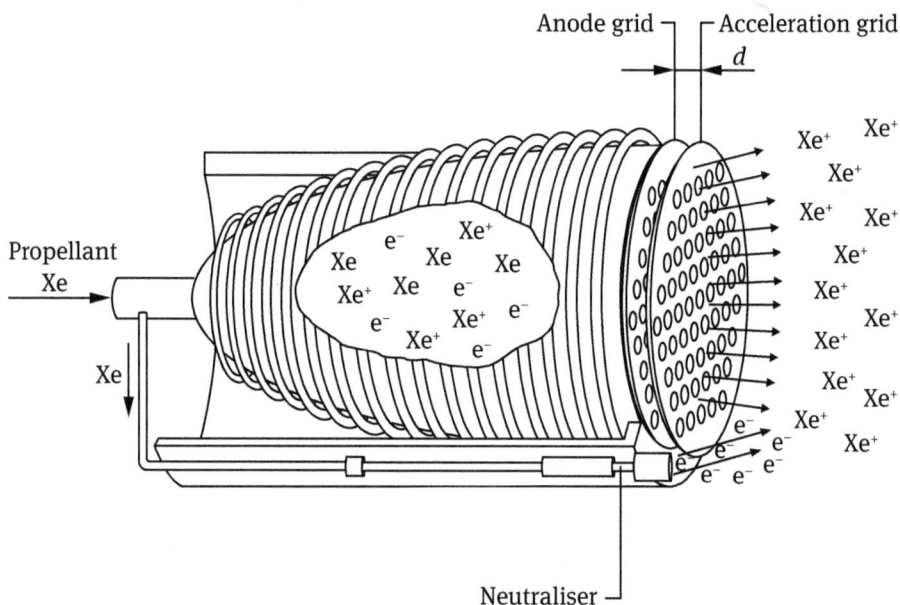

Figure 3.10: Principle design of an electrostatic engine with radio-frequency ionisation.

Therefore, the limitation of the space charge decreases with increasing length of the distance of acceleration d (from the anode grid to the accelerating grid), which means that this distance has to be kept very small. The diameter of electrostatic thrusters has to be chosen large enough to realise maximum ion current, thus the mass of the thrusters is relatively large compared with chemically propelled thrusters.

Table 3.6: Characteristics of some typical electrostatic thrusters.

Type	Propellant	I_{sp} / s	Power / W	Thrust / mN
Field emission thrusters	Caesium	6 000	275	5
Kaufman thrusters	Mercury or Xenon	2 500	4 000	200
RIT thrusters	Mercury or Xenon	3 200	400	10

Table 3.7: Properties of propellants for electric thrusters.

Propellant	Atomic no.	Atomic mass	Ionisation energy / eV	Ionisation energy/ Wh·kg^{-1}
Xenon (gas)	54	131.3	12.1	2470
Caesium (solid)	55	132.9	3.9	790
Mercury (liquid)	80	200.6	10.4	1390

It is expected that electric thrusters are increasingly being used for future attitude control of commercial satellites, e.g. for north-south station keeping (NSSK) with GEO satellites. Also for interplanetary missions with high velocity demand (mercury missions, missions to comets, etc.) electric thrusters are required with high specific impulse and thrusts of up to 1 N.

Several electrostatic thrusters are currently under development for these tasks. Some requirements and properties of the RIT-XT thruster of Airbus DS are presented in Table 3.8.

Table 3.8: Requirements and properties of the RIT-XT thruster. Reference: Airbus DS GmbH.

Feature	Value	Property	Value
Specific impulse	3000 s – 4500 s	Beam diameter	220 mm
Thrust	>150 mN	Nominal operating voltage	2200 V
Power consumption	<4500 W	Grid distance	<1 mm
Propellant consumption	<20 g·h^{-1}	Anode grid	
Mass	<8.5 kg	Grid voltage	2000 V
		Wall thickness	<0.5 mm
		Bore diameter	<2 mm
		Material	Molybdenum
		Accelerating grid	
		Grid voltage	– 200 V
		Wall thickness	<1.5 mm
		Bore diameter	<1.5 mm
		Material	Graphite
		Operating pressure	≈ 10^{-6} mbar
		Period of operation	>10 000 h

Low operating pressure and long period of operation require enormously technical efforts, in particular on testing technology. Electrostatic thrusters work more effectively if less foreign particles can be hit by Xenon ions. Since charge changes always occur, charged particles are no longer focused. They are then attracted by the grids where they cause processes of erosion which reduce the operational life of the thrusters. For a correct ion-optical focusing of the particles and an unhindered passage through the holes of the grid an adjusted operational voltage has to be applied to the grids for various operating points.

3.4.3 Thermonuclear Propulsion Systems

Thermonuclear propulsions utilise the binding forces of protons and neutrons in atomic nuclei. In principle, the thermonuclear propulsion systems can be grouped to the type of energy generation into four categories.

– Natural radioactive decay
– Nuclear fission
– Thermonuclear fusion
– Annihilation of matter

It is an energy source with extremely high energy densities which is basically interesting for space applications. For propulsion purposes, similar to chemical propulsion, the propellant is heated by these energy sources via a heat exchanger, the energy source is cooled at the same time, and subsequently used as recoil mass by thermodynamic relaxation as with chemical propulsions. Depending on the temperature limit of the energy source, exhaust velocities up to $10\,000\,\mathrm{m\cdot s^{-1}}$ can be achieved with hydrogen (molecular weight $2\,\mathrm{g\cdot mol^{-1}}$) as propellant.

Natural radioactivity

Radioactivity is a property of radionuclides, so-called isotopes, of certain chemical elements. Unstable atomic nuclei of these isotopes decay spontaneously. During this nuclear reaction the nuclei of the radioactive isotope decay into nuclei of a lighter chemical element emitting radiation as high-energetic alpha or beta particles. In many cases, the fission products of the radioactive decay are in an exited state and gamma radiation is released after further nuclear decay of these products. The radiation released is converted to thermal energy after being absorbed by substances.

In the field of nuclear physics, a distinction is made between naturally produced radioisotopes and artificially radioisotopes produced by nuclear reactions. Radioisotopes of elements after bismuth, atomic number 83, in periodic table of the elements mostly have short half-lives. Today, they cannot be detected if they would not continually created anew from the radioactive decay of the long-lived uranium isotopes uranium-238 and uranium-235, and thorium-232 as intermediate product of a decay series.

Besides the element uranium, some lighter elements are used for radioactive age determination of the formation age of minerals, rocks, glasses, and other samples. Artificial radioisotopes are created by human-made nuclear reactions. They are created by neutron irradiation in nuclear reactors or by irradiation with charged particles in particle accelerators. Artificially generated radioisotopes are used in radiotherapy with dubious success. But dramatic effects of radioactive radiation on biological matter are verifiable. As soon as the amount of radiation exceeds a certain limit the germination capacity of seeds is reduced. Growth and developmental retardation and malfunctions in animals including humans are observed. Mammalian tissues are even more sensitive to radiation whenever they divide fast, e. g. gonads and blood-forming cells.

Every radioisotope has a characteristic half-life, in which half of the number of initially existing atomic nuclei decay. Thus, long-lived isotopes have a lower energy density as short-lived isotopes. The radioactive decay of atomic nuclei is already being used in today's space applications for onboard power supply of satellites by means of radioisotope thermoelectric generators (for RTGs see Chapter 5.4). The use for propulsion purposes failed, due to the weight of the installations with relatively low thrust and environmental compatibility. There is a considerably higher risk of radioactive contamination during lift-off.

Nuclear fission

The energy output can be considerably increased if nuclear fission is achieved by neutron bombardment instead of spontaneous natural nuclear fission. This means that an atomic nucleus of uranium-235 releases an energy of 200 MeV. This corresponds to an energy density of $22.8 \cdot 10^9 \, \mathrm{Wh \cdot kg^{-1}}$ of uranium-235. A nuclear reactor is operated with a thermal output of 30 kW per kilogramme enriched uranium. This corresponds to approximately one hundred times the energy density of a radioisotope thermoelectric generator (RTG) with natural radioactive decay.

There are many different types of reactors and processes to ensure energy supply for the whole mankind and future generations for the long term. Risks of application and handling (e. g. nuclear reprocessing) are so enormous by the high risk potential, such as toxicity, exposure to radiation as the result of the war, terrorist attacks, and also operating errors, that nuclear energy will account only for a part of the future energy supply in the long term. The disposal of radioactive waste and safe storage over millennia is a long-term burdensome inheritance for future generations, which lead also to very controversial opinions among scientists concerning the ethical responsibility of technology.

Thermonuclear fusion

The energy output can be further increased if, instead of nuclear fission of heavy atomic nuclei, thermonuclear fusion of light atomic nuclei is used. During fusion of one gramme of hydrogen to helium-4 an energy of approximately $600 \cdot 10^9 \, \mathrm{J}$ is released. This corresponds to an energy density of $166 \cdot 10^9 \, \mathrm{Wh \cdot kg^{-1}}$ of hydrogen. In principle, thermonuclear fusion has similar risk potentials like nuclear fission and acceptance problems thereby. Solutions to extensive technical problems of the use of thermonuclear fusion for the peaceful purpose of supplying energy are currently not in sight for an unforeseeable time, because of incalculable development risks along with corresponding financial needs. Application potentials for propulsion technology of spacecrafts still appear a long way off from implementation.

Annihilation of matter

The highest theoretically possible energy density is achieved by annihilation of matter. The complete annihilation of the rest mass according to the equivalence of mass and energy

$$E = m \cdot c^2 \qquad (3.20)$$

postulated by *Albert Einstein* in 1905/1907 results in $90\,000 \cdot 10^9$ J per gramme of matter. This corresponds to an energy density of $25 \cdot 10^{12}$ Wh$\cdot$kg^{-1} of matter. The Equation 3.17 for the equivalence of mass and energy must not be confused with the equation for kinetic energy of inertial mass

$$E = \frac{1}{2} m \cdot v^2 \qquad (3.21)$$

In the event of a complete annihilation, matter combines with antimatter. Large deposits of antimatter in the universe are not currently aware. However, annihilation of matter takes place on a large scale in the cosmos near black holes if masses are being accelerated close to the speed of light. Annihilation of matter on Earth takes place on the submicroscopic scale in particle accelerators for research purposes. Using annihilation of matter for terrestrial energy generation is not expected in the foreseeable future. Also for astronautics this source of energy is currently not available due to technical problems and it is mentioned here only for the sake of completeness.

3.4.4 Photon Propulsion Systems

Photon propulsion systems are mere hypothetical procedures, with which the impulse of a focused photon beam is used for generating thrust. Highly simplified, the light beam of headlights is comparable with a photon propulsion. However, a sufficient high thrust would only result from very intensive photon radiation, such as that emitted by radiation sources with temperatures above 30 000 K. The radiation maximum of a photon emitter enters the area of soft x-ray radiation with increasing temperature. Focusing of the photon beam is then possible only by grazing reflection, analogous to the Wolters telescope in astronomy with parabolic or hyperbolic dishes. Then, a conventional reflection is impossible. The energy P for generating a thrust of 1 N is $P = F \cdot c = 300$ MW.

There are no acceptable proposals for realisation of photon propulsions so far. This propulsion procedure is not used for space flight for an unforeseeable time. *Eugen Sänger* intensively examined this issue scientifically. Because of the interesting theoretical possibility to reach other solar systems and galaxies this propulsion was very popular with early space flight enthusiasts as well as science-fiction authors.

Solar sailing

In contrast to all other propulsion systems the necessary recoil mass (in general propellants) need not to be carried by the satellite during solar sailing, but is supplied by light particles of solar radiation. According to the Theory of Special Relativity every energy corresponds to a certain fictitious mass based on $E = m \cdot c^2$.

The change of direction of this mass causes a counter reaction analogous to all other propulsion systems. During reflection the incident beam and the reflected beam lay in the same plane together with the perpendicular of incidence and form equal angles. Angle of incidence is the angle of reflection. As the intensity of solar radiation decreases with the square of the distance to the source of radiation (here: the Sun), the quality of this propulsion system is strongly dependent on distance. The thrust can be calculated according to the following equation.

$$F = (1 + \eta) \frac{S}{c} \tag{3.22}$$

Where

F is the radiation pressure
η is the degree of reflection ($0 < \eta < 1$)
c is the speed of light
S is the local solar constant

In the realm of the Earth ($S = 1372\,\text{W} \cdot \text{m}^{-2}$) the thrust is up to a maximum of $9\,\text{N} \cdot \text{km}^{-2}$ at total reflection. This low thrust involves relatively large sails and their correct unfolding after launch from Earth is a major problem. Moreover, these sail areas increase the total mass of the satellite because both the mirrored foils and the superstructure have a mass of approximately $10\,\text{g} \cdot \text{m}^{-2}$. For comparison, the mass of a typical sheet of paper is $80\,\text{g} \cdot \text{m}^{-2}$. The following figure shows the resulting effect of different sail positions.

The sails are from plastics (e. g. Kapton, Mylar) of a thickness of approximately $1\,\mu\text{m} = 10^{-3}\,\text{mm}$. The side facing to the Sun is coated with a reflective layer which typically may have a reflectivity up to a maximum of 90 %. Solar sailing can be carried out for an unlimited period of time analogous to electric propulsion systems resulting in spiral orbital transitions. High supply velocities can be realised despite of low thrusts because of the long operational period. The conditions mentioned above suggest that a commercial use of this propulsion system is not to be expected in the foreseeable future. Even without intended solar sailing the radiation pressure of the Sun constitutes a disturbance of the free-flight trajectories of satellites and needs to be corrected regularly by another propulsion system depending on the mission profile.

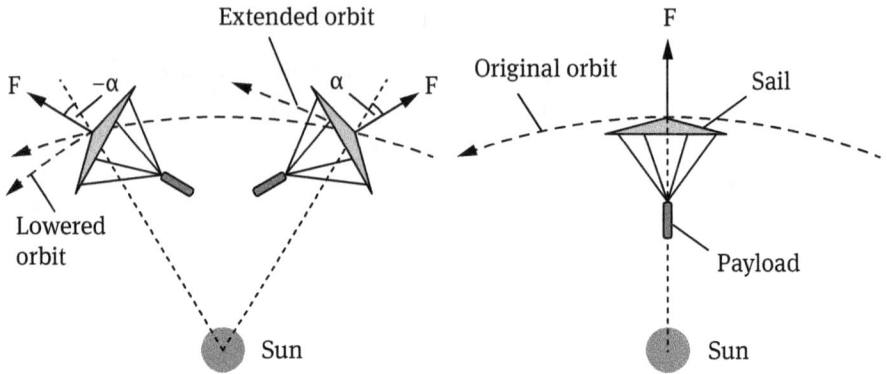

Figure 3.11: Solar sail positions.

The sails are from plastics (e.g. Kapton, Mylar) of a thickness of approximately $1\,\mu m = 10^{-3}\,mm$. The side facing to the Sun is coated with a reflective layer which typically may have a reflectivity up to a maximum of 90%. Solar sailing can be carried out for an unlimited period of time analogous to electric propulsion systems resulting in spiral orbital transitions. High supply velocities can be realised despite of low thrusts because of the long operational period. The conditions mentioned above suggest that a commercial use of this propulsion system is not to be expected in the foreseeable future. Even without intended solar sailing the radiation pressure of the Sun constitutes a disturbance of the free-flight trajectories of satellites and needs to be corrected regularly by another propulsion system depending on the mission profile.

3.5 Questions for Further Studies

1. Give the names of different propulsion systems used in space applications.
2. What does the exhaust velocity of the combustion gases of a de Laval nozzle depends on?
3. Give advantages and drawbacks of solid propelled propulsions.
4. What does the exhaust velocity of an electrostatic thruster depends on?
5. Give advantages and drawbacks of electric propulsion systems.
6. What thermonuclear propulsion systems are available for space applications?
7. Inform yourself about nuclear rocket projects in past and present.
8. Give advantages and drawbacks of sun sailing.
9. Explain the physical principles of solar radiation pressure.
10. What is the minimal power of a spotlight (e.g. torch) being operated to generate a thrust of 1 N?

4 Missions

Every celestial body is surrounded by a gravitational field (see Chapter 2). To leave the gravitational field physical work must be performed. This means that energy is required. The physical situations of our astronomical environment are the result of the following energy states E in the gravitational field of individual celestial bodies.

Table 4.1: Energy states.

Celestial body / moon	E_{total} / kWh·kg^{-1}	ΔE_{total} to Earth / kWh·kg^{-1}	E_{total} (planet) / kWh·kg^{-1}
Sun	− 52966	− 52843	− 52966
Mercury	− 318.3	− 195.1	− 2.5
Venus	− 170.3	− 47.1	− 14.9
Earth	− 123.2	± 0	− 17.3
Earth's Moon	− 0.144	+ 17.2	− 0.78
Mars	− 80.9	+ 42.3	− 3.5
Ceres	− 44.5	+ 78.7	− 0.046
Jupiter	+ 23.7	+ 99.5	− 492.2
Io	− 41.7	+ 141.2	− 0.91
Europa	− 26.2	+ 125.7	− 0.57
Ganymede	− 16.4	+ 115.9	− 1.05
Callisto	− 9.3	+ 108.8	− 0.83
Saturn	− 12.9	+ 110.3	− 174.8
Titan	− 4.3	+ 114.6	− 0.97
Uranus	− 6.4	+ 116.8	− 62.9
Neptune	− 4.1	+ 119.1	− 76.6
Triton	− 2.7	+ 121.8	− 0.29
Pluto	− 3.1	+ 120.1	− 0.21

The first column 'E_{total}' in Table 4.1 describes the position of planets and moons in the gravitational field of the superordinated centre of gravity. The value for the Sun refers to an orbit close to its surface. The total energy also corresponds to the kinetic energy around the respective centre of gravity. To leave the Solar System from an Earth-like orbit an energy of 123 200 Wh·kg^{-1} is required. Only 144 Wh·kg^{-1} are required to leave the gravitational field of the Earth from a Moon-like orbit.

Column 'ΔE_{total}' shows the differences of these energy states relative to our home planet Earth. Among others, this results in a considerable more difficult accessibility of the inner planet Mercury (195 100 Wh·kg^{-1} for a permanent orbit) than of the outer dwarf planet Pluto. The nearest inner planet Venus and the nearest outer planet Mars are relatively easy to reach.

Column 'E_{pot} (planet)' describes the size (or better depth) of the gravitational field of individual planets and moons. The values each refer to the surface of the celestial body. To leave the gravitational field of the Earth an energy of 17 300 Wh·kg^{-1} is therefore required while leaving the gravitational field of the Moon only 780 Wh·kg^{-1} are required.

4.1 Velocity demand

Due to energy-related reasons, the minimum requirements in respect of velocity for different missions can be calculated by first approximation according to the principles from Chapter 2. An eventually existing atmospheric drag is not taken into account, as well as course corrections. The velocity demand required is shown in the two following tables.

Remarks on the following tables

Column 1: Mission with flight destination and transitional locations for propulsion manoeuvres.

Column 2: Typical flight time (with an asterisk * as propulsion phases. See column 5, otherwise free-flight phase).

Column 3: Altitude (local altitude above the surface of the celestial body).

Column 4: v_∞ / m·s^{-1} (velocity at infinity relative to the celestial body).

Column 5: $v_{location}$ / m·s^{-1} (velocity relative to the celestial body).

Column 6: v_{kick} / m·s^{-1} (velocity demand required for propulsion manoeuvres).

Column 7: v_{loss} / m·s^{-1} ('x' stands for unspecified, not negligible velocity losses, (aerodynamic drag, gravitational ascent losses, etc.).

Column 8: v_{total} / m·s^{-1} (sum of the propulsion manoeuvres from column 6) since the beginning of the mission).

Column 9: Comments

Realised = technically realised

Not realised = technically not realised at present (possible in principle)

Possible = technically possible at present

Not possible = technically not possible at present

Near Earth destinations are listed in Table 4.2. More than 90 % of all space missions remain in the gravity field of the Earth, including the International Space Station (ISS) in the low Earth orbit (LEO), which is the biggest space project at present. Currently, approximately 15 % of all space missions, which are being launched annually, reach the ISS as destination. A final velocity of 7730 m·s^{-1} is needed for approaching the ISS. For the ascend of the launcher another velocity demand (see Table 4.2, column 7, v_{loss}) of approximately 1000 to 1200 m·s^{-1} must be taken into account. Space applications in the middle Earth orbit (MEO) are becoming increasingly important. Mainly various and future satellite constellations for a worldwide broadband access to the Internet (e. g. OneWeb) or publicly available Earth observations (e. g. Skybox) use these orbit heights.

Sun-synchronous polar orbits (SSO) also belong to the middle (extended) Earth orbits (MEO). There are increasing requirements for these orbits at the end of life (EOL) of a satellite to clear the orbit by de-orbiting and a controlled descend of the satellite.

Table 4.2: Near Earth destinations (see also remarks on page 82).

Mission	Flight time	Altitude	V_∞	$V_{location}$	V_{kick}	V_{loss}	V_{total}	Comment
To LEO	15 min*	300 km		7730	7730	x	7730	Realised
From LEO	5 min*	300 km		7730	90		7820	Realised
Back from LEO	44 min	0 km		0				
To MEO	15 min*	300 km		8022	8022	x	8022	Realised
At apogee	51 min	1414 km		6874			8022	Realised
At MEO	5 min	1414 km		6874	281		8303	Realised
FromMEO	5 min	1414 km		7155	367		8670	
Back from MEO	49 min	0 km		0				
To GEO	20 min*	300 km		10157	10157	x	10157	Realised
At apogee	5 h 16	35800 km		1607			10157	Realised
At GEO	10 min*	35800 km		1607	1468		11625	Realised
From GEO	10 min*	35800 km		3074	1499		13124	Not realised
Back from GEO	5 h 13	0 km		0				
To the Moon	20 min*	300 km		10839	10839	x	10839	Not realised
At apogee	5 d 2 h	384E3 km	825	185			10839	
At Moon's orbit	10 min*	100 km		2452	819		11658	
To the surface	1 min*	100 km		1633	23		11681	
At peri-luneum	57 min	0 km		1702			11681	
Soft landing	5 min*	0 km		1702	1702	150	13533	
Launch	5 min*	0 km		0	1702	150	15385	
At apo-luneum	57 min	100 km		1610			15385	
Within Moon's orbit	1 min*	100 km		1610	23		15408	
From Moon's orbit	10 min*	100 km	829	1633	821		16229	Possible
Back from Moon	5 d 2 h	0 km		11096				
Synodic period	24 h 49 min							
To the Moon in 3 days	20 min*	300 km		10865	10865	x	10865	Realised
At Moon's altitude	3 d	384E3 km	1112	769			10865	
At Moon's orbit	10 min*	100 km		2563	930		11795	
Landing (see above)					3450	300	15545	
From Moon's orbit	10 min*	100 km	1114	1633	931		16476	
Back from Moon	3 d	0 km		0			16476	Apollo
Escape Earth parabolic	20 min*	300 km	0	10931	10931	x	10931	Realised
At Moon's altitude	2 d 4 h	384E3 km	0	1429				

Navigation satellite applications like the GPS satellites operated by the 50th Space Wing (50 SW) of the U.S. Air Force and the Galileo global navigation satellite system (GNSS) of ESA are placed into orbits of approximately 20000 km above Earth. However, the greatest commercial significance have geostationary orbits (GEO) and geosynchronous orbits (GSO). The GEO is used for meteorological satellites, communication satellites, and satellites for TV transmission (e. g. Astra or Eutelsat). Approximately 25 % of all space missions are placed into GEO.

Table 4.3: Planets and stars (see also remarks on page 82).

Mission	Flight time	Altitude	v_∞	v_location	v_kick	v_loss	v_total	Comment
To Venus	20 min*	300 km		11232	11232	x	11232	Realised
At Earth's level		150E6 km	2582	27095			11232	
At Venusian level	21 w	107E6 km	2809	37913			11232	
Into orbit of Venus	10 min*	300 km		10497	3345		14577	
At peri-centre	45 min	0 km		7415	87		14664	
Soft landing	30 min	0 km		0			14664	Realised
Launch	10 min*	0 km			0	7415 3000	25079	Not possible
At apo-centre	45 min	300 km		7064			25079	
From Venusian orbit	10 min*	300 km		7151	3345		28511	
At Venusian level		107E6 km	2809	37913			28511	
At Earth's level	21 w	150E6 km	2582	27095			28511	
Back from Venus		0 km		0			28511	
To Mars	20 min*	300 km		11254	11254	x	11254	Realised
At Earth's level		150E6 km	2676	32352			11254	
At Mars level	36 w	220E6 km	2432	22091			11254	
Into orbit of Mars	10 min*	300 km		5395	1990		13244	
At peri-centre	53 min	0 km		3627			13317	
Soft landing	20 min	0 km		3627		10	13327	Realised
Launch	10 min*	0 km		0	3627	700	17654	Not realised
At apo-centre	53 min	300 km		3332			17654	
From orbit of Mars	10 min*	300 km		3405	1990		19717	
At Mars level		220E6 km	2432	22091			19717	
At Earth's level	36 w	150E6 km	2676	32352			19717	
Back from Mars		0 km		0			19717	Possible
Into orbit of Mars	10 min*	300 km	0	5395	579		11833	Mariner
Into orbit of Venus	10 min*	300 km	0	10497	383		11615	Magellan
Into orbit of Mercury	15 min*	100 km	0	10572	6407		19717	Messenger
Into orbit of Jupiter	10 min*	10000 km	0	56075	282		14282	Galileo
Into orbit of Saturn	10 min*	10000 km	0	33259	444		15442	Cassini
Into orbit of Uranus	10 min*	10000 km	0	18575	594		16262	Not realised
Into orbit of Neptune	10 min*	10000 km	0	20287	408		16352	Not realised
Leaving Solar System	20 min*	300 km	12292	16450	16450	x	16450	Realised
At Moon's altitude	8 h 37 min	384E3 km	12292	12375				
At Earth's level		150E6 km		41968				
At Jupiter	1 a 6 w	780E6 km		18439				Voyager
At Saturn	2 a 31 w	145E7 km		13526				Voyager
At Uranus	6 a 38 w	285E7 km		9649				Voyager
At Neptune	13 a	450E7 km		7679				Voyager
To Alpha-Centauri	30 min*	300 km	20236	23000	23000	x	23000	Maximum
At Earth's level		150E6 km	27016	49912			23000	
Jupiter swing-by	36 w	10000 km	31442	32709	38457		61457	
At Jupiter's level		780E6 km	39813	43875			61457	
Saturn swing-by	41 w	10000 km	37719	42048	20711		82168	
At Saturn's level		145E7 km	43593	45677			82168	
At Alpha-Centauri	28894 a	4 2 ly	43593	43593			82168	
$v_\text{rel.} = -25\,\text{km}\cdot\text{s}^{-1}$	18369 a	2 7 ly	68593	68593			82168	
Into the Sun	20 min*	300 km		28021	28021	x	28021	Not possible
At Earth's level		150E6 km	25801	3875			28021	
At Sun's surface	9 w 3 d	600E3 km		450619				

Example for an interplanetary flight to Venus

Some distant astronomical destinations are listed in Table 4.3. We launch for Venus from Earth's surface with a 20-minute propulsion manoeuvre of the launch vehicle and reach a hyperbolic orbit relative to Earth at an altitude of 300 km with a local velocity of 11 232 m·s^{-1}. During this ascent manoeuvre significant losses of velocity x, typically approximately 2 000 m·s^{-1}, occur.

Relative to the Sun, we have an altitude of Earth's orbit of 150E+6 km and a velocity of 27 095 m·s^{-1}. This is equivalent to a velocity v_∞ of 2 582 m·s^{-1} relative to Earth. After 21 weeks, we arrive at Venus at an altitude of 107E+6 km towards the Sun with a local velocity of 37 913 m·s^{-1}. This corresponds to v_∞ of 2 809 m·s^{-1} relative to Venus. To reach a Venusian orbit at an altitude of 300 km, a manoeuvre of 10 minutes must be carried out by 3 345 m·s^{-1} to reduce velocity. This reduces the local velocity in the peri-centre from 10 497 m·s-1 to 7 152 m·s^{-1}.

With another manoeuvre to reduce velocity of 87 m·s^{-1} the altitude of the peri-centre is reduced to 0 km. We now reach pendulously on parachutes the surface of Venus without another braking manoeuvre. For all propulsion manoeuvres, we used *in summa* a required velocity of 14 664 m·s^{-1}. By comparison, Apollo required approximately 16 476 m·s^{-1} to the Moon and back.

Launch from the Venusian surface is technically not possible at present because a required velocity of at least 7 415 m·s^{-1} is necessary for orbit. During ascent significant atmospheric drag occurs (approximately 3 000 m·s^{-1}). Incidentally, we would get along with a pressure of 100 bar at an ambient temperature of 500 °C, so that this astronomical destination is considered as being unreachable for manned missions. For return to Earth in our example, a propulsion of 28 511 m·s^{-1} is required by calculation.

Example for minimum requirements for interplanetary flights

To reach a planet or moon from Earth, a space probe has to be injected at least into a parabolic orbit around the planet or moon. The braking manoeuvre (at the peri-centre, column 3) for a transition from the hyperbolic approach path into a parabolic orbit is shown in column 6. The total energy required is shown in column 8. A relatively great expense is striking at Mercury. This problem (Hohmann transfers) is avoided by the use of electrical propulsions for continuous use and spiraling to approach Mercury's orbit (planned mission of ESA: BepiColombo).

Example for interstellar destinations

To leave our Solar System with the today's propulsion technologies and the use of swing-by manoeuvres a theoretical maximum velocities of 45 km·s^{-1} is possible at best. By comparison, in 2015 the fastest New Horizons space probe so far passed the dwarf planet Pluto with a flyby velocity of 13.5 km·s^{-1}.

Flight times in the rough order of magnitude of 100 000 years are necessary to reach solar-near exoplanets, such as the red dwarf *Gliese 832* at a distance of 16 light years. Our neighbouring solar system, *Alpha Centauri*, moves relatively towards us with a velocity of $25 \, \text{km} \cdot \text{s}^{-1}$, and therefore enabling a shortened travel according to Table 4.3 within 18 369 years.

Example for near sun mission

To fly from the Earth into the Sun nearly the total velocity of the inhabited spacecraft Earth must be decelerated. During a flight time of nine weeks a perihel of 600 000 km could thus be reached and the total mass of the mission would evaporate by the solar heat. Such a mission is not possible with the today's propulsion systems. The European Space Agency (ESA) has commissioned a solar-near mission (Solar Orbiter) with a perihel of 0.3 AU (45 000 000 km).

4.2 Questions for Further Studies

1. What historical and current space missions do you know?
2. Why planet Mercury is more difficult to reach to perform scientific investigations of its surface than the dwarf planet Pluto?
3. Give the reasons why planet Pluto has been classified as a dwarf planet.
4. What is the problem of escape from a planet by a rocket?
5. Define the terms 'apo-luneum' and 'peri-luneum'.
6. What is the purpose of the celestial coordinate system?
7. By which numbers a point is specified within a spherical coordinate system?
8. What measurements have been carried out to determine if and when the Voyager 1 space probe reached the interstellar medium?
9. Give the extend of the solar system.
10. Which objects orbit the Sun directly and which indirectly?

5 Energy Sources

In this chapter the sources for energy supply of on-board systems of spacecrafts are described. The energy sources of propulsion systems were already described in Chapter 3. Every satellite has an electric energy supply for instruments und control units. The board voltage is approximately 30 V direct current. Because there are no standards, applications from 16 V (satellite constellation Globalstar) up to more than 70 V are well-known. The potentially usable energy sources can be roughly divided into three groups.

- Chemical fuels
    - for thermodynamic energy supply (thermal combustion).
    - for electrochemical energy supply (cold burning).
- Nuclear fuels
    - for nuclear fusion.
    - for nuclear fission.
- Solar cells

Steam power plants for energy supply

In steam power plants the steam is generated in the boiler, then relaxed in a turbine and liquefied in the condenser. The condensing temperature is selected to be high, that the heat to be removed is being able to be used for heating purposes, e.g. within a district heating systems (power-heat coupling). In large-scale power plants the waste heat is removed unused by cooling towers or disposed of with the river water.

The terrestrial output of steam power plants is more than 80 % of the worldwide electrical energy produced. For this purpose fossil fuels (hard coal, lignite, crude oil, natural gas, peat), nuclear fuels (uranium, plutonium), and regenerative heat sources (wood, straw, thermal solar energy) serve as a source of heat. For space applications these facilities are eliminated because power density and reliability is too low for a large number of mechanically moving parts.

In space technology special energy sources are often used to ensure specific requirements concerning high energy density and reliability which are not used in a terrestrial environment for cost efficiency reasons. Typical examples of applications are solar cells and radioisotope thermoelectric generators (RTGs) for unmanned missions and fuels cells for manned missions. Meanwhile, the production costs for solar cells and fuel cells were reduced to a range which permits an economically viable deployment for terrestrial applications, such as commercial vehicle drives or cogeneration units.

Comparative examples

To perform an estimation and evaluation of the energy density for individual energy sources and energy storages some comparative figures are noted here. One watt-hour (Wh) is equivalent to an energy of $3600\,\text{Ws} = 3600\,\text{J}$. An energy of $4200\,\text{J}$ is required to warm up 1 litre of water by $1\,°\text{C}$. That is

$$\frac{4200\ \text{J}}{\text{kg} \cdot 3600\ \text{s}} = 1.2\,\text{Wh}\cdot\text{kg}^{-1}$$

This energy corresponds to a height difference (e. g., waterfalls) of

$$m\cdot g\cdot h = 4200\text{J} \;\Rightarrow\; h = 400\,\text{m}$$

or to a translational velocity of

$$\frac{1}{2}m\cdot v^2 = 4200\ \text{J} \;\Rightarrow\; v = 335\,\text{km}\cdot\text{h}^{-1}$$

- Evaporation of 1 litre of water requires $627\,\text{Wh}\cdot\text{kg}^{-1}$.
- Ionisation of 1 kg Xenon requires $2470\,\text{Wh}$.
- Combustion of hydrogen and oxygen to 1 litre of water with cooling to $25\,°\text{C}$ provides $4400\,\text{Wh}\cdot\text{kg}^{-1}$.
- Combustion of coal by oxygen provides a combustion energy of $2500\,\text{W}\cdot\text{kg}^{-1}$ per 1 kg of carbon dioxide.

To move a satellite from Earth's surface into an orbit at least $8650\,\text{Wh}\cdot\text{kg}^{-1}$ are required. Exactly double energy of $17\,300\,\text{W}\cdot\text{kg}^{-1}$ is required to kick the satellite out of the gravitational field of the Earth, as independent artificial celestial body around the Sun.

The Earth has a kinetic energy of approximately $123\,000\,\text{Wh}\cdot\text{kg}^{-1}$ on its orbit around the Sun. Exactly the same energy is required to kick a satellite into its Earth-like orbit out of the gravitational field of the Sun, as independent artificial celestial body in the Milky Way on a sun-like orbit.

In 2013, the annual energy consumption on Earth was approximately 18.2 billion tonnes HCU (primary energy consumption (hard coal, lignite, crude oil, natural gas, nuclear energy, renewable sources of energy) by private households, traffic and industry). A hard coal unit (HCU) is equivalent to a combustion energy of 1 kg hard coal. It is defined as $7000\,\text{kcal} = 29.3\cdot10^6\text{J} = 8141\,\text{Wh}$. This results in a global energy consumption of $148\cdot10^{15}\,\text{Wh}$. It is equivalent to a power of $17\,000\,\text{GW}$.

5.1 Batteries

Batteries are electrochemical (galvanic) elements, in which a chemical reaction takes place spatially separated in two separate reactions at current carrying electrodes (anode and cathode). In this process, the energy of formation of the reactants is directly converted into electrical energy with high efficiency.

The negative electrode is referred to as anode und the positive electrode is referred to as cathode. Within the electrolyte the positive anions move towards the anode and the negative cations move towards the cathode.

Galvanic elements with an unidirectional reaction process are designated to as primary systems. In different galvanic elements chemical reactions can also be reversed by reversing the polarity of the electrodes. This property is used for energy storage in secondary cells, so-called accumulators. The two batteries of the moon buggy were such a secondary system (silver-zinc). They were used as primary system without recharging.

Table 5.1: Typical primary systems.

Name Anode / Electrolyte / Cathode	Voltage / V	Energy density / $Wh \cdot kg^{-1}$
Dry battery ('Lechlanché cell')	1.5 – 1.6	80
Zn / NH_4Cl / MnO_2		
Alkaline dry cell battery	1.3	100
Zn / KOH / MnO2		
Zinc-mercury battery (button cell)	1.6	110
Zn / KOH / HgO		
Lithium battery	1.85	300 – 500
Li / org. or anorg. electrolyte / SO_2 or $SOCl_2$		

5.2 Fuel Cells

In contrast to the galvanic elements described above, fuel cells are electrochemical power sources, in which the reactants are supplied continuously. Reaction products have to be removed. In an alkaline fuel cell (AFC) an aqueous potassium hydroxide solution is used as electrolyte. Hydrogen gas is conveyed to the anode and subsequently oxidised to H^+. At the cathode hydroxide ions (OH^-) are generated by reduction of oxygen. Together with the H^+ water (H_2O) is formed at the anode, which has to be continuously led away as predominant reaction product. The operating temperature of an alkaline fuel cell is 60 °C to 120 °C.

A fuel cell with a strong alkaline electrolyte which is supplied by hydrogen and oxygen is illustrated in Figure 5.1. The secondary formation of hydrogen peroxide, other possible secondary reactions, and the proportion of the alkali metal ions in the electric current are not shown.

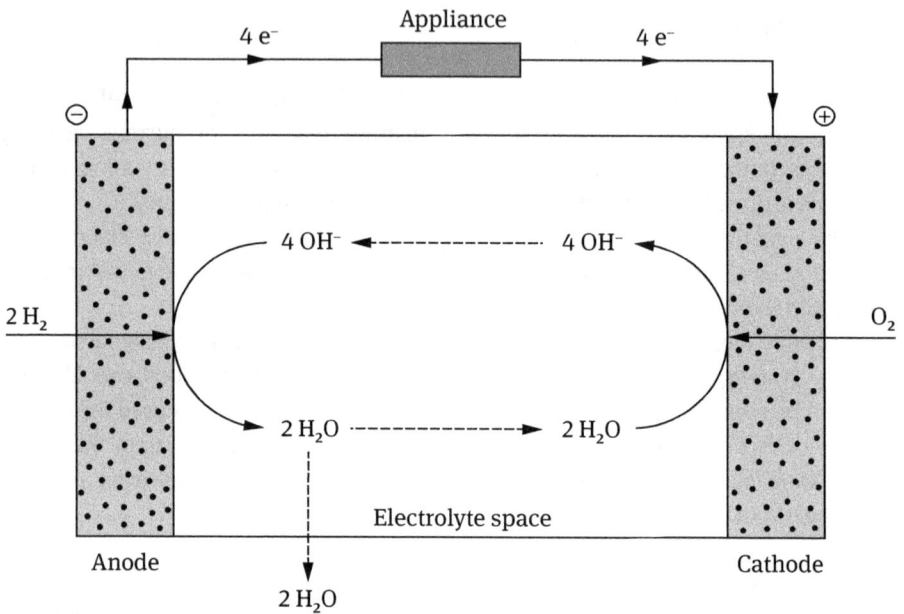

Figure 5.1: Reaction scheme of an alkaline fuel cell.

There are different types of fuel cells indicated by abbreviations. The principle of operation is the same for all. Table 5.2 below shows three different types of fuel cells used in aerospace applications.

Table 5.2: Typical fuel cells.

Name Anode / Electrolyte / Cathode	Voltage / V	Abbreviation	Energy density / $Wh \cdot kg^{-1}$ (at long-term discharge)
Alkaline fuel cell H_2 / KOH / O_2	1.2	AFC	120 (compressed-gas operation) 120 (cryogenic operation, Apollo)
Acidic fuel cell H_2 / H_3PO_4 / O_2	1.23	PAFC	120 (compressed-gas operation)
Hydrazine fuel cell N_2H_4 / KOH / O_2	1.56	HFC	700

Fuel cells require minimal maintenance. They are non-polluting, have a favourable partial load behaviour, a high power to weight ratio, and a high degree of efficiency. For the emerging use of fuel cells for electric drives and terrestrial energy supply electrolytes with molten carbonates and ceramic electrolytes are being used in the high-temperature range. In the long term, the aim of ecologically harmless closed-loop energy recycling can be realised by fuel cells. However, the costs of developing reliable and safe systems have not yet been calculated by detail.

Thermodynamics of fuel cells

Within open systems the Gibbs free energy (also known as free enthalpy), named after the U.S. American physicist and mathematician *Josiah William Gibbs*, is the maximum work profitable during reversible isothermic changes of state. In fuel cells the Gibbs free energy ΔG of the combustion reaction is converted into electrical energy, often designated to as 'cold burning'. Due to the internal resistance and inhibition of the electrodes losses occur. Due to the cell reaction

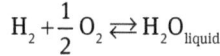

$$H_2 + \frac{1}{2}O_2 \rightleftharpoons H_2O_{\text{liquid}}$$

the enthalpy of formation (calorific value) is $\Delta H = -285.25\,\text{kJ}\cdot\text{mol}^{-1}$ at $298\,\text{K}$. The Gibbs free energy is $\Delta G° = -236.6\,\text{kJ}\cdot\text{mol}^{-1}$. This allows the maximum thermodynamic or ideal efficiency of

$$\eta_{th} = \frac{\Delta G°}{\Delta H} \tag{5.1}$$

$$\eta_{th} = \frac{-236.6}{-285.25} = 0.829$$

to be defined. In this case, heat is released into the surrounding environment by the fuel cell. For $\eta_{th} > 1$ heat is removed from the surrounding environment by the fuel cell (heat pump) as it is the case of the cell reaction

$$C + \frac{1}{2}O_2 \rightleftharpoons CO \text{ with } \eta_{th} = 1.98$$

with

$$U_{00} = -\frac{\Delta G°}{n\cdot F} \tag{5.2}$$

the standard electromotive force U_{00} (standard EMF) can be calculated. Where n is the number of electrons generated per reaction, here $n = 2$ electrons, and F is the Faraday constant ($96\,484\,\text{As}\cdot\text{mol}^{-1}$), which is the electrical charge of 1 mole of electrons.

$$U_{00} = \frac{236\,600}{2\cdot 96484} = 1.23\,\text{V}$$

This open-circuit voltage is a theoretical value which drops to measured $0.95\,\text{V}$ at room temperature and ambient pressure.

Method of operation of fuel cells

More specifically, combustion reactions take place in fuel cells. The idea to keep on delivering electric energy from a galvanic cells by means of electrochemically active substances for an unlimited period of time is attributed to *William Robert Grove* who engineered hydrogen-oxygen fuel cells in 1839. Modern works started around 1950 and boomed as galvanic systems with high energy densities were sought for space applications.

In an alkaline electrolyte the transport of charge is mainly achieved of up to 75% by hydroxide ions, OH^-. The rest is achieved by ionic migration of potassium ions, K^+. Sourcing reactions occur as

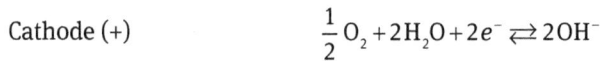

Anode (−) $\qquad\qquad\qquad\qquad H_2 + 2H_2O^- \rightleftarrows 2H_2O + 2e^-$

Cathode (+) $\qquad\qquad\qquad\qquad \frac{1}{2}O_2 + 2H_2O + 2e^- \rightleftarrows 2OH^-$

The anode, on which hydrogen is fed, is here negative, because an oxidation take place. At the positive cathode a reduction proceeds. In the case of an acidic electrolyte as phosphoric acid, H^+-ions are the most important charge carriers. Sourcing reactions occur as

Anode (−) $\qquad\qquad\qquad\qquad H_2 \rightleftarrows 2H^+ + 2e^-$

Cathode (+) $\qquad\qquad\qquad\qquad \frac{1}{2}O_2 + 2H^+ + 2e^- \rightleftarrows 2H_2O$

Within the electrolyte the negative anions migrate to the negative anode driven by a concentration gradient and the positive cations migrate to the positive cathode. This gives rise to the names anode and cathode.

Gas diffusion electrode

Gas diffusion electrodes are used as electrodes which form a three-phase zone as large as possible, in which the gas phase, the liquid phase, and the solid phase adjoin to each other (Figure 5.2). Gas diffusion electrodes must essentially meet the following three requirements. For all three cases gas electrodes become inoperative.

–   Electro-catalytically high active to achieve highest possible current densities.

–   Pores of the gas electrode must not be soaked with electrolyte by means of high capillary forces.

–   Electrolyte must not be completely displaced from the pores by means of high gas pressure.

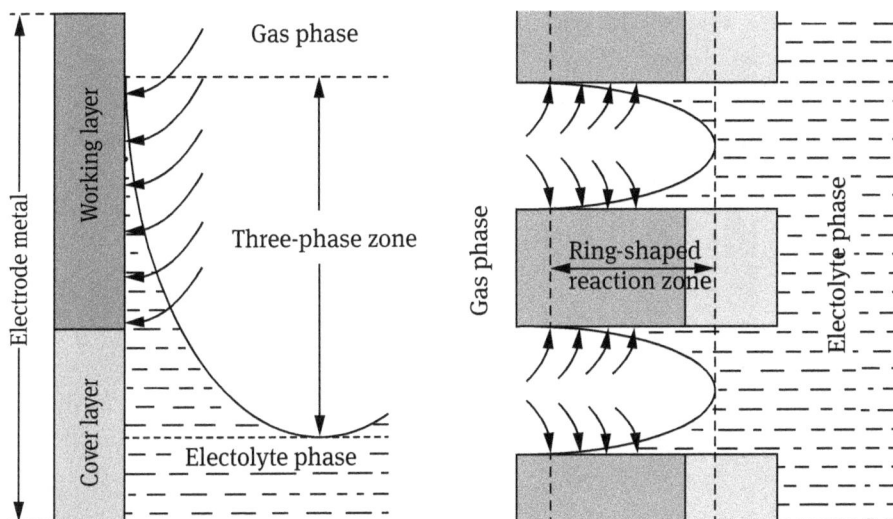

Figure 5.2: Three-phase zone of a gas diffusion electrode[4].

In particular, platinum metals and their alloys are suitable for conversion of hydrogen, and also special nickel preparations such as Raney nickel which is cheaper. Silver is also suitable for conversion of oxygen. The so-called electro-catalysts are embedded in the porous walls of suitable substrate materials (e. g., sintered nickel or compressed activated carbon) in small quantities.

Today, most fuel cells for space applications work in the range of atmospheric pressure up to some bar overpressure and from approximately 50 °C to over 180 °C. The state of the art shall be illustrated by the examples of the Apollo fuel cells in Figure 5.3 (left) with aqueous electrolyte and the fuel cells used in the Space Shuttle with absorbed electrolyte in Figure 5.3 (right). The older Apollo cells were operated at a working temperature of 193 °C to 238 °C and at a pressure of 4.1 bar The electrolyte was aqueous 75% potassium hydroxide solution above 100 °C. The removal of the reaction water was carried out via a hydrogen circuit. Water is formed at the anode and evaporated into the hydrogen stream. Each module was hosted in a pressure jacket for temperature control. Inside the pressure jacket nitrogen circulated to remove the heat of reaction via a radiant cooler, which is the reason for a high operating temperature. For heating during initiation an embedded heating coil was used. 31 cells were assembled to a module (1.12 kW at 28 V with a mass of 110 kg). Three modules generated an electric energy of up to 500 kWh with an overall weight of 810 kg during a 10-day flight. That means, a specific energy density of 620 Wh · kg^{-1} was achieved.

---

4 Hamann CH, Hamnett A, Vielstich W. Electrochemistry. Wiley-VCH Verlag, Weinheim, 2007.

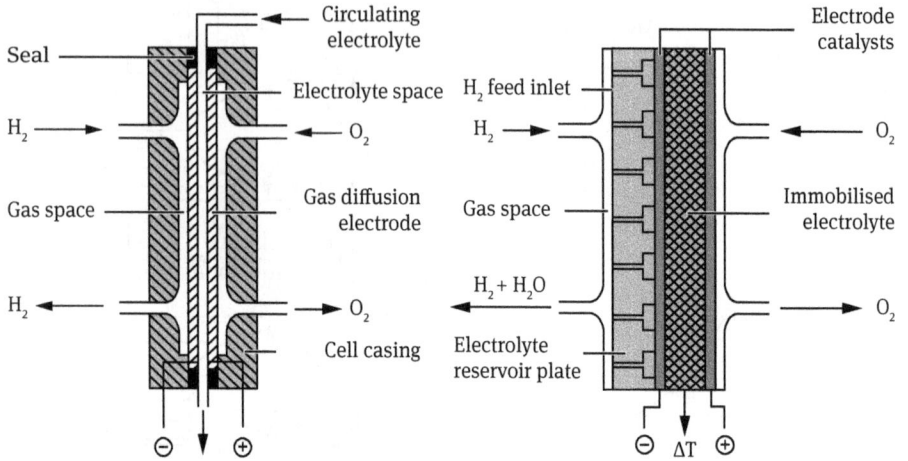

Figure 5.3: Basic design of fuel cells used in the Apollo capsule (l.) and in the Space Shuttle (r.).

There are three units in the Space Shuttle (see Figure 5.4). The peak load is 12 kW at 27.5 V and the minimum load is 2 kW at 32.5 V. Each unit consists of three parallel batteries, each with 32 cells coupled in series. The working temperature is 88 °C to 108 °C at an operating pressure of 4.1 bar. The electrolyte consists of 35 % potassium hydroxide solution. The caustic solution has its conductivity maximum at this concentration and is immobilised in a matrix. The reaction water is drained via hydrogen circuit by an condensing heat exchanger and separator. There is a separate cooling circuit for heat dissipation.

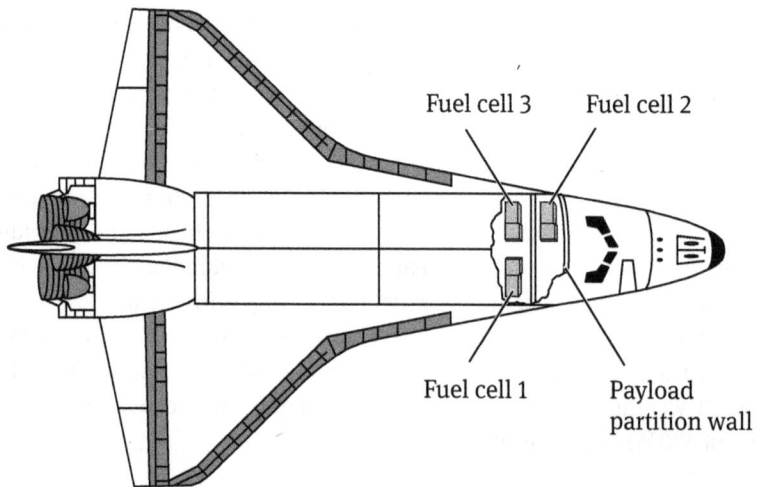

Figure 5.4: Fuel cells in the forward position of the orbiter's mid-fuselage.

5.3 Solar Cells

Direct conversion of solar energy into electric energy is achieved by solar cells. In the year 1954 a method for the production of solar cells was discovered by the Bell Telephone Laboratories in the USA. A thin silicon platelet coated with a much thinner layer of silicon saturated with Boron converted the sunlight impinging upon it directly into electricity. A large number of photovoltaic cells are used to build up a solar cell. Incident photons of the light cause a separation of positive and negative charges in these cells. The voltage generated impels the so-called photon-induced current by which the cell produces its power. This leads to an increase of the energy yield by an increasing intensity of the solar radiation. A number of solar cells are assembled to solar batteries or solar generators. Best solar cells are made of monocrystalline silicon or gallium arsenide with an efficiency of over 30%. In the past, solar cells were rigid panel components. Today, cost-effective manufacturing of solar cells in the form of lightweight and flexible foils is possible.

Originally developed for space flight, solar cells were used at first in the field of entertainment and consumer electronics. Today, solar cells are used mainly as power sources in the area of mobile appliances. Traffic guidance systems and lightning systems on motorways are supplied by solar cells, and some public street lightnings.

Throughout the world, there are a number of research projects with solar power plants, solar cars, and solar airplanes. Currently, the construction is uneconomical because of excessively higher investment costs. By progressive development in semiconductor technologies and decreasing production costs in solar technologies the generation of electricity by lage-scale of solar cells becomes more attractive. The development, production, and the use of solar cell panels, associated with the related fields of semiconductor technology and solid-state chemistry, are referred collectively to as photovoltaics. For space applications solar generators can be evaluated as follows.

- The solar constant has to be calculated. In the realm of the Earth it is approximately $1370 \, W \cdot m^{-2}$.
- Considering the efficiency of a solar power plant the maximum electric power can be calculated to approximately $200 \, W \cdot m^{-2}$.
- Depending on the type of the satellite (spin stabilisation or three-axis stabilisation) the average exposure of the solar panels to the sunlight can be calculated by continuous tracking of the solar panels towards the Sun.
- From the average power requirements of the satellite the surface area of the solar generators can be calculated.

Today, the power requirements of larger communication satellites are already over 10 kW. Solar generators with panels of over $100 \, m^2$ are already in use for permanent and almost unlimited energy supply.

During launch of a satellite, solar generators are folded and transported in the payload fairing of the launch vehicle. It is always a special challenge to unfold the solar generators after deployment of the satellite into space. Solar cells are temperature sensitive. In the GEO orbit typical equilibrium temperatures are at $+60\,°C$ in the sunlight and at $-180\,°C$ in the shadow of the Earth[5]. This load and other environment conditions cause a reduction of the efficiency over time. Always at the end of a mission (EOL) only a lower current output is still available.

Generating electricity from solar cells for interplanetary missions is only of interest within the range of the inner planets because of the dependency on the intensity of sunlight. For missions beyond the distance of Mars other energy sources have to be used (e. g., radioisotope thermoelectric generators).

5.4 Thermonuclear Energy Sources

The highest energy densities of energy sources that have been used so far are achieved by radioisotope thermoelectric generators at present. In the so-called RTGs permanent heat release is generated by radioactive materials through nuclear fission.

Plutonium-238 is normally used as plutonium dioxide for space flight. It is an alpha-radiation source (helium particle) with a half life of 89 years and requires only little radiation protection in contrast to beta and gamma radiators. The thermal energy released by nuclear fission is 390 W per kilogramme of plutonium-238. Due to its high toxicity special security precautions must be made when working with or operating RTGs. In the United States the use of RTGs as energy source is subject to an approval procedure which simulates every conceivable technical failure. The approval procedure is closed by the President of the United States.

If there is no space for large solar generators or fuel cells which would require an extensive infrastructure, RTGs may also be used in the near-earth realms. Even the lunar lander of the Apollo project used RTGs of type SNAP-27. The electrical power was 75 W (thermal output of 1500 W) at 16 V direct current and a mass of 20 kg of one RTG. The radioisotope used was plutonium-238 of 3.8 kg. In April 1970, after the Apollo 13 accident, such a capsule together with the lander splashed down in the Pacific Ocean. The capsule remained intact[5].

A higher energy density can be theoretically provided by nuclear fusion. The Sun uses nuclear fusion for energy production and thereby enables life on our planet. On Earth, nuclear fusion of light isotopes (hydrogen, deuterium and tritium) is currently at an early stage of development. Today, thermonuclear reactors are stable only in the millisecond range. Still more energy has to be used to ignite the

5 Ruppe HO. Introduction to Astronautics. Vol. 1 / Vol. 2. Academic Press, London, 1966/1967.

fusion fire than fusion energy can subsequently be recovered. Because of the extreme ambient conditions (high temperatures in the range of several million Kelvin and extreme low pressures) the development costs for cost-effective use cannot be reliably estimated at this time. The terrestrial use of nuclear fusion achieved a macabre fame as weapon of mass destruction (hydrogen bomb). The fusion process is initiated by ignition of a fission nuclear bomb and energy release caused thereby, which show a significant increase compared with 'conventional' fission nuclear bombs. This energy source is not utilisable for space applications with existing techniques far into the future.

The Galileo Jupiter probe had two RTGs with the sum of 570 W at the beginning of the mission (BOL) and 480 W at the end of the mission (EOL). The Cassini Saturn probe has three RTGs with the sum of 33 kg of plutonium. For more than 20 missions beyond the distance of Mars, RTGs with 100 % reliability have been used up to now (Voyager, Viking, Ulysses, and others).

5.5 Thermoelectric Modules

For direct conversion of thermal energy into electric energy and *vice versa*, two already well-known effects from the 19th century are used. In 1821, the Baltic German physicist *Thomas Johann Seebeck* discovered the thermoelectric effect, the direct conversion of temperature differences to electric voltage. Some years later the French phyisicist *Jean Charles Athanase Peltier* discovered a reverse effect in 1834. Today's thermoelectric transducers, which generates power by a temperature difference, are called Peltier elements.

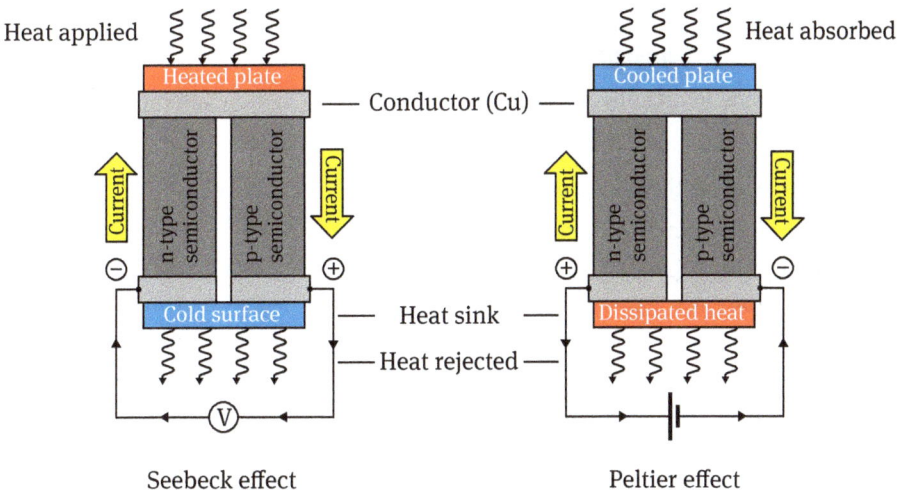

Figure 5.5: Thermoelectric modules (TEM).

Peltier effect

If an interface of two different conductors is passed through by a current, one of the contact sides cools down while the other one heats up. With some semiconductors, this effect is relatively large, so that a technical application is worthwhile for cooling and heating purposes within a Peltier element. The reversal of this thermoelectric property is called Seebeck effect.

Seebeck effect

An electric voltage (thermoelectric voltage) is created between two spatially separated points of contact of two different materials, if there is a temperature difference between them. Thermoelectric generators with an efficiency of more than 10 % can be manufactured by the selection of materials that have relatively high electrical conductivity and low thermal conductivity as possible. Within these generators pairings of p- and n-doped semiconductors with thermoelectric voltages of up to $500\,\mu V \cdot K^{-1}$ are used. In space flight, semiconductor devices composed of PbTe/ZnSb[6] are frequently used. They have a thermoelectric output voltage of $180\,\mu V \cdot K^{-1}$.

5.6 Questions for Further Studies

1. What energy sources of on-board energy supply systems of spacecrafts do you know?
2. Give advantages and drawbacks of radioisotope thermoelectric generators.
3. Why are fuel cells not used for energy supply of satellites?
4. Why are lithium batteries increasingly used for energy supply?
5. How does the thermoelectric technology work?
6. How can the Peltier effect work?
7. Name the three laws of thermoelectricity.
8. How does solar thermal energy get converted into electricity?
9. What is the meaning of thermal electricity?
10. What, in any case, does 'thermal' mean?

---

6 Lead-telluride-zinc-antimonide

6 Energy Storages

In the event that sources of energy are not permanently available (e. g., the satellite passes through the Earth's shadow), an intermediate storage of energy is necessary during the period of power failure. This power failure could also be caused by sudden, unforeseen events. For example, a short circuit which may arise in connection to high radiation exposure at the time of sunspot maxima. For this case, an emergency power supply must have reliable access to suitable energy storages.

6.1 Mechanical Flywheels

Electric energy can be stored in the rotating mass of mechanical flywheels as rotational energy. For charging, the polarity of the current generator is reversed and acts powering as an electric motor on the flywheel mass. Torques occur during charging which have to be compensated by the propulsion system.

The energy storage density is limited to approximately $10\,\mathrm{Wh \cdot kg^{-1}}$ by the mechanical strength of the parts. Here, numbers of revolutions of over 30 000 rpm are necessary. Residual friction of the magnetic bearing flywheel masses and moving parts reduce reliability, and thereby the use is limited to special applications in space flight. Flywheels are being used with increasing frequency for attitude stabilisation of rotationally symmetric satellites (see Chapter 2.4.2).

6.2 Electrochemical Storages

Electrochemical storages are rechargeable galvanic elements (accumulators and secondary elements). To supply satellites with energy, a combination of solar cells to convert energy and storage batteries has become established. Photovoltaic solar cells are used as energy converters to convert light energy into electrical energy. The electrical energy is then stored in sealed and maintenance-free storage battery systems. A long lifespan is the most important characteristic.

Storage battery systems are selected according the requirements which arise from the use of the specific satellite. GEO satellites mostly have nickel-cadmium accumulators. They bridge downtimes of the solar generators (maximum of 70 min) when the satellite passes through the Earth's shadow during the equinox in spring or autumn. Meanwhile, nickel-metal hydride accumulators with a d. c. voltage of 1.5 V, and lithium-ion batteries with a d. c. voltage of 3.6 V – 3.7 V have been developed for environmentally friendly terrestrial applications without cadmium.

Table 6.1: Typical secondary systems.

Name Anode / Electrolyte / Cathode	Voltage / V	Energy density / Wh·kg^{-1}	Typical number of cycles with low discharge level
Lead-acid accumulator Pb / H$_2$SO$_4$ / PbO$_2$	2.06	25 – 35	10 000
Nickel-cadmium accumulator Cd / KOH / NiOOH	1.30	30 – 40	7 000
Silver-cadmium accumulator Cd / KOH / AgO	1.60	60 – 70	4 000
Silber-zinc accumulator Zn / KOH / AgO	1.85	100	100
Na-S accumulator (350 °C) Na / Na$_2$O · 11Al$_2$O$_3$ / S	2.10	100	1000

6.3 Chemical Propellants

Above all, mass and volume related energy density and storage life are particularly important for the use of chemical propellants as energy storage. The highest energy densities occurs during oxidation of light metals (Li, Be, B, Mg, Al) with oxygen, ozone, and fluorine. Due to difficult handling and toxicity only systems are generally used with a lower energy density. The storage of cryogenic propellants (hydrogen, methane, and oxygen) is suitable for this purpose. The storage life is limited to several weeks or months due to the loss of heat insulation and the resulting boil-off rate. A long-term storage of propellants is possible at room temperature. Hydrazine, its derivatives MMH, UDMH, and hydrocarbons can be stored over years.

6.4 Questions for Further Studies

1. What energy storage systems of space applications do you know?
2. Give advantages and drawbacks of lead-acid accumulators.
3. What form of energy is stored in a mechanical flywheel?
4. Give other important characteristics of battery systems used for satellites?
5. What is the reason for the different voltages supplied by secondary systems?
6. What is the difference between secondary cells and primary cells?
7. Give the reason for the different charge cycles of secondary systems.
8. What do the abbreviations MMH and UDMH stand for?
9. Which components the Aerozine 50 propellant consists of?
10. Give the reason for the long shelf life of hydrazine and hydrocarbons

7 Materials and Lubricants

To meet the specified requirements of the environmental conditions (see Chapter 12) for different space products, the materials used must have properties which are only of minor importance to some extent in earth-based applications. Therefore, the number of the types of materials is very large and a more detailed description would exceed the scope of this book.

The universe consists to 75 % of hydrogen and to 24 % of helium. The remaining portion consists of about 100 different chemical elements which exist in increased concentrations on Earth. Three quarters of the elements are metals, the remaining part being noble gases (He, Ne, Ar, Kr, Xe, Rn), halogens (F, Cl, Br, J, At), elements of living nature (C, H, O, N, S, P), and seven semi-metals (B, Si, Ge, As, Sb, Se, Te). For space applications the following properties are application-specific mattered.

7.1 Mechanical properties

In general, light metals with high strength (light metal alloys of titanium, aluminum, and magnesium) and composites (GRP, CFRP, C/SiC, etc.) are used for stressed mechanical components. The strength can be further increased by design using special honeycomb structures to save even more weight.

- Material characteristics
    - Density
    - Tensile strength
    - Yield strength
    - Compressive strength
    - Fatique strength
    - Cycling strength
    - High-temperature strength
    - Endurance strength
    - Torsional strength
    - Modulus of elasticity
    - Modulus of shearing
    - Poisson's ratio
    - Hardness
    - Dynamic viscosity
    - Kinematic viscosity
    - Porosity

- Chemical properties
    - Stoichiometric valency
    - Corrosion behaviour
    - Enthalpy of formation (combustion energy)
- Thermal properties
    - Thermal conductivity
    - Thermal expansion coefficient
    - Heat capacity
    - Vapour pressure
    - Melting point
    - Boiling point
    - Triple point
    - Critical point
    - Adiabatic coefficient
    - Heat of fusion
    - Heat of evaporisation

- Optical properties
    - Light emission coefficient
    - Reflection coefficient

- Electrical and magnetic properties
    - Resistivity
    - Ionisation energy
    - Transition temperature (for superconductors)
    - Critical current density (for superconductors)
    - Debye temperature (for superconductors)
    - Magnetic susceptibility
    - Thermoelectric voltage
    - Electrochemical voltage

- Physical properties
    - Atomic or molecular weight
    - Half-life (with radio isotopes)
    - Radiation energy (with radio isotopes)
    - Cross section (e.g. for neutrons)
    - Sputtering rate (e.g. with lattices for electrostatic engines)

- Biological properties
    - Toxicity (e.g. maximum workplace concentration)

7.2 Lubrication properties

The properties of lubricants under space conditions are of particular importance. The term tribology is used for an own branch of science which concerns itself with the investigation of friction processes. Many of the known lubricants are not suitable for use under space conditions. Soft metals (lead, silver, gold, etc.) are often used or chemical compounds with high vapour pressure (MoS_2, Teflon (PTFE)).

7.3 Materials Used in Space

Titanium

Titanium and titanium alloys offer the highest potential of weight savings for spacecraft vehicles. Therefore, titanium alloys had already been applied earlier in Apollo and Mercury capsules. Classical areas of application of titanium alloys are sheet metals for propellant tanks of rockets and satellite tanks. Titanium sheet metals are formed superplastically to hemispheres and simultaneously diffusion welded. Low weight, high strength and chemical long-term tolerance against the propellants are of great importance. The special titanium alloy Ti-3Al-2.5V was developed for low temperature applications. Due to its good toughness and ductility down to cryogenic temperatures, this alloy was used for high-pressure tubes of the hydrogen pumping system of the Space Shuttle.

Aluminum

Since the early days of space flight aluminum has been used for all types of space structures. Chosen for its light weight and the ability to withstand stresses during launch and operation in outer space, aluminum has been used for the Apollo spacecraft, the space laboratory Skylab, the Space Shuttle and the International Space Station. Due to its high volumetric energy density and the difficulty to ignite accidentally, aluminum was used as the primary propellant for the SRB motors of the Space Transportation System. The primary structures of NASA's Orion Multi-Purpose Crew Vehicle are made from aluminum-lithium alloy and will be covered by an improved version of thermal protection tiles used on the Space Shuttle.

Magnesium

Magnesium and magnesium alloys are the lightest structural metal or alloys used in aerospace. Magnesium has a density of $1.8\,g \cdot cm^{-3}$. By volume magnesium is 30% lighter than aluminium and 75% lighter than steel. Magnesium alloys are flexible and versatile materials which can be shaped into sheet, plated by rolling, casted using sand, extruded into both solid and hollow profiles, and processed as powder and granules. These different material states can be further shaped using forging, pressing, folding, and machining from the solid material.

Composites

A composite material is defined as a material system which consists of a mixture or combination of two or more micro-constituents which are mutually insoluble and differing in form and/or material composition. Examples of composite materials are fiber reinforced plastics (ceramics plus polymers), vinyl-coated steel (metals plus polymers), and steel reinforced concrete (metals plus ceramics). Composites are versatile used for both structural applications and components to achieve weight reduction, thermal stability, high impact resistance, and high damage tolerance.

7.4 Commodity Prices

The commodity costs for the materials used are often only of minor importance, so that even more expensive precious metals (gold, platinum, rhodium, iridium, rhenium, etc.) or artificially generated radioactive isotopes are widely used.

It must be mentioned for gross cost accounting that the today's transportation costs of 1 kg of payload into Earth's orbit are between 10 000 and 100 000 euros. In general, the mass-specific costs for actual payloads are even more expensive. Today, a big surveillance satellite costs one billion euros or 50 000 euros per kilogramme. Thus, the mass-specific costs are therefore higher than the commodity prices of the most expensive precious metals.

7.5 Questions for Further Studies

1. What is the principal difference between the environmental conditions of outer space and those on Earth?
2. What is the influence of those environmental conditions on the selection of suitable materials?
3. What properties play a role for selecting materials?
4. Explain the term tribology in more detail.
5. What are composites in general?
6. What are ceramic matrix composites in particular?
7. What do the abbreviations GRP and CFRP stand for?
8. Name the differences between C/SiC and C/C-SiC composites?
9. Where is Teflon used in aerospace technology?
10. What is the coefficient of friction?

8 Processes

The materials used have to be subjected to a variety of procedures and processes for manufacturing suitable space products. In the following only MAIT processes (**M**anufacturing, **A**ssembly, **I**ntegration and **T**est) should be considered. The processes for product development and project-processing as well as business processes are not discussed. Moreover, no operational processes are addressed in this chapter (e. g. fuelling, ignition, launch, controlling, adjusting, evaluation, etc.).

After a thorough and critical design review the production activities begin with manufacturing and assembly. All production activities include assembly processes, system integration processes, and verification processes. Thus, the processes of manufacturing, assembly, integration and testing can be divided into

– production processes,
– verification processes, and
– testing processes.

Integration takes place in a special room with defined conditions such as temperature and air humidity, vibration damping, electromagnetic compatibility, dust-free conditions, etc. The system integration process is closely associated with the system verification. Among others, the objectives of the system verification of space products are

– design qualification,
– proof and confirmation of the design qualification,
– accuracy of manufacturing,
– proof of the performance parameters, and
– confirmation of the qualification of the mission staff.

A wide range of space products are manufactured by the processes listed above. Further processes are not listed here. The products to be processed can be divided into the following main groups, depending on extent and complexity with increasing functionality.

– Piece parts
– Assemblies
– Components
– Subsystems
– Systems

8.1 Manufacturing processes

A great number of manufacturing processes are specific for space applications. The following list shows some examples.

- Chipling
    - Turning
    - Milling
    - Drilling
        - Mechanical
        - Laser
        - Electrochemical
    - Grinding
        - Machine sanding
        - Honing
        - Polishing
            Manually
            Machine-polishing
            Electropolishing
        - Lapping
    - Carving
    - Filing
    - Splitting
        - Sawing
        - Cutting
        - Laser cutting
        - Scraping
    - Threading
- Chipless machining
    - Extrusion (extrusion moulding)
    - Compression moulding
    - Calendaring
    - Bending
    - Crimping
    - Rolling
    - Punching
    - Spinning
    - Pulling
    - Pressing
    - Transfer moulding
        - Forging
        - Pruning

- Casting
- Injection moulding
- Electrochemical processing
  - Galvanising
  - Eroding
- Chemical processing
  - Pickling
  - Etching
- Thermal processing
  - Cooling
  - Drying
  - Annealing
  - Hardening
  - Quenching
  - Sintering
  - Glassing
  - Flame cutting
- Coating
  - Painting
  - Varnishing
  - Anodising
  - Passivation (zinc coating, etc.)
- Processing of masses
  - Milling
  - Mixing
  - Filtering
  - Pelletising
  - Kneading
  - Loading (e. g. catalyst for monopropellant thrusters)
- Joining
  - Assembly
    - Screws and screw retention
    - Pins and rivets
    - Crimping
    - Bonds
    - Press-fits
    - Shrinking
    - Snap fastenings

- Joining (continued)
  - Thermal joining
    Welding
    Fusion welding
    - Gas welding
      - Arc welding
        - Metal arc welding
        - Inert gas welding
          - Tungsten inert gas (TIG)
            - TIG orbit and TIG manually
          - Submerged arc welding
      - Beam welding
        - Electron beam welding and laser welding
    Press-fit welding
    - Arc stud welding and explosion welding
    - Ultrasonic welding and friction welding
    - Resistance press-fit welding
      - Spot welding and projection welding
      - Seam welding and flash welding
    Soldering and brazing
    Environment (vacuum, inert gas, ...)
    Temperature (high-temperature, low-temperature, ...)
    Method (hard soldering, soft soldering, ...)
- Cleaning
  - Brushing
    - Wiping
    - Blowing-off
    - Suctioning
    - Ultrasound cleaning
    - Megasound cleaning
    - High-pressure cleaning
    - Watery cleaning
    - Chemical cleaning
    - Low-pressure plasma cleaning
    - CO_2 cleaning
    - Laser cleaning
  - Labeling
    - Laser
    - Etching
    - Bonding
    - Manually

8.2 Verification Processes

The processes of manufacturing inspection for quality assurance are listed below in this chapter. The manufactured components are tested by different processes.

- Dimensional inspections[7]
    - Mechanical
    - Manually
    - By machine
    - Optical
- Visual inspections
- Metallographic inspections
- Non destructive inspections
    - Ultrasonic inspection
    - X-ray inspection
    - Gamma ray inspection
    - Dye penetrant inspection
    - Eddy current inspection
    - Magnetic resonance tomography
- Hydraulic tests
    - Pressure difference / flow rate
    - Optical
- Static tests
    - Tensile test
    - Pressure test
    - Bending test
    - Torsion test
    - Hardness test
    - Surface inspections
- Dynamic tests
    - Impact bending test
    - Fatigue
    - Hardness test

---

7 Length, angle, radii, shape, etc.

8.3 Testing Processes

This chapter describes the processes for function testing. Tests in the propulsion mechanism area can be divided into cold tests and hot firing tests. The manufactured products are tested application-specifically in different processes for reliable functioning.

Cold tests

- Leakage test (external / internal)
- Pressure test
- Burst test
    - Non-destructive
    - Destructive
        - Flow tests
    - Gas
    - Water
    - Test liquid
    - Propellants

- Electrical functional tests
    - Open and closing response
    - Resistance and impedance
    - Power
    - Pull-in, drop-out
        - Environmental tests
    - Vibration
        Sinus
        Random
    - Shock test
    - Acoustic noise
    - Thermal vacuum tests
    - Climate tests
- Cleanliness verification

Hot firing tests

- Conditions at sea level
- Vacuum (up to 0.1 mbar)
- Operating modes
    - Continuous operation[8]
    - Pulse-mode firing
- Functional tests
    - Performance[9]
    - Flow rate
    - Temperatures
    - Pressure[10]
    - Vibration
- Decontamination[11]

---

8 Steady state
9 Specific impulse
10 Static and dynamic
11 Hypergolic propellants

8.4 Test philosophy

Load factor and extent of the tests are specified within the way of performing the test, depending on complexity, criticality, and designated use of a space product.

Figure 8.1: Large-scale test of the main engine of the Ariane 5 rocket during of the development of the 1000 kN thrust chamber at Lampoldshausen. Test duration: max. 20 s, test rates: one test per week. Image: ©Airbus DS GmbH.

Thus, different test requirements and loads are defined for the following successive products and models.

- Pre-development models (PDM)
- Development models (DM)
- Pre-qualification models (PQM)
- Qualification models (QM)
- Pre-flight models (PFM)
- Flight models (FM)

Loads of a space product can be

- proof pressures,
- operating loads,
- operating time, and
- environmental loads (vibration, shock, etc.).

Before delivery to the customers, flight models (FM) are subjected to acceptance tests which neither have any damaging impact nor reducing the lifetime of the product. In general, pre-flight models encounter higher loads while still remain flight worthy.

The necessary justification for the specified requirements concerning the space product is demonstrated on qualification models with higher loads. These models are not flight worthy anymore at the end of testing and no longer used in space missions.

8.5 Questions for Further Studies

1. Describe the way of doing tests and verifications in the acceptance of space products.

2. What are the typical loads which space products are exposed?

3. What are the minimum requirements for electrical connectors used in any space environment?

4. Which properties have to be considered in material selection for a space station?

5. Which conditions could contribute to the deterioration of hardware?

6. Which metal should not be used in crew environments of a space station at temperatures above 100 °C?

7. Which particular influences materials are exposed in low Earth orbit?

8. Explain the technology of rapid prototyping as integral part of the development process.

9. Give the advantages of the rapid prototyping technology.

10. Which techniques are used in rapid prototyping?

9 Products

Depending on the field of application, highly reliable components and systems are required for performing tasks during space flight. Requirements for these products are defined by specifications and proved by extensive tests. The following list of typical components and subsystems conveys an impression of the abundance of the components.

- Life support systems for manned missions
- Propulsion systems for altitude and orbit control (AOCS) with
    - Propellant tanks
    - Gas tanks
    - Pressure regulation units with
        Valves
        Pressure regulators
        Propellant filters and gas filters
    - Fill & drain valves
    - Propulsion systems with
        Valves
        Pumps and turbines
        Gas generators
    - Transducers for temperature, pressure, vibration, etc.
- Mechanical structures of subsystems, instruments, and thrust frames
- Thermal hardware
    - Isolation of tanks and liquids
    - Thermal protection of re-entry bodies
- Energy supply and storage
    - Solar generators
    - Radioisotope thermal-electric generators (RTGs)
    - Fuel cells and batteries
- Instruments for scientific, commercial, and military tasks
- Antennas
- Navigation systems
- On-board computers and software

There is a small number of reliable manufactures for each product worldwide. The reliability of the complex overall system depends on the reliability of all individual components and related criticality in case of individual components' failure, contrary to specification requirements. To some extent, manufactures are partly hindered from supplying sensitive technology to customers of the global market by trade barriers such as export controls and the like. Some systems applied are exemplified below.

9.1 Launch Vehicles

Each space mission starts with launch preparations of the payload launcher system. Exemplarily, four systems are presented.

- Saturn V (lunar rocket of the 1960s and 1970s)
- Ariane 5 (manufactured by Airbus DS and operated by Arianespace)
- Space Shuttle (re-usable manned launcher system, in service until 2011)
- Space Launch System (launch planned before November 2018)

9.1.1 Saturn V

The Saturn V rocket was developed at Marshall Flight Center in Huntsville, Alabama, USA. Two smaller rockets, the Saturn I and Saturn IB, were used as predescessors. These smaller rockets were used to launch humans into Earth's orbit. The first Saturn V was launched in 1967. The mission was called Apollo 4 followed by Apollo 6 in 1968. These two rockets were launched without crews and tested the rocket. On Apollo 8 mission a crew was launched into space for the first time.

The Saturn V lunar rocket was a three-staged launcher system for transporting payloads into transfer orbit to reach the Moon. The first stage of this rocket was the most high-thrust and most powerful reliable propulsion system ever built, hence enabling flights to the Moon. Table 9.1 gives some characteristics of the rocket.

Table 9.1: Characteristics of a Saturn V rocket.

Characteristic	First stage	Second stage	Third stage
Mass, complete	2135 t	458 t	115.2 t
Mass, propellant	2000 t	422 t	104.5 t
Mass, dry	127 t	32.6 t	9.5 t
Propellants	LOX / RP-1 (kerosene)	LOX / LH2	LOX / LH2
Engines	5 x F-1 (Rocketdyne)	1 x J-2 (Rocketdyne)	1 x J-2 (Rocketdyne)
Thrust (5 engines)	3400 t (s. l.), 3875 t (vac.)	1 x 90.8 t (vacuum)	1 x 90.8 t (vacuum)
Firing time	150 s	400 s	500 s

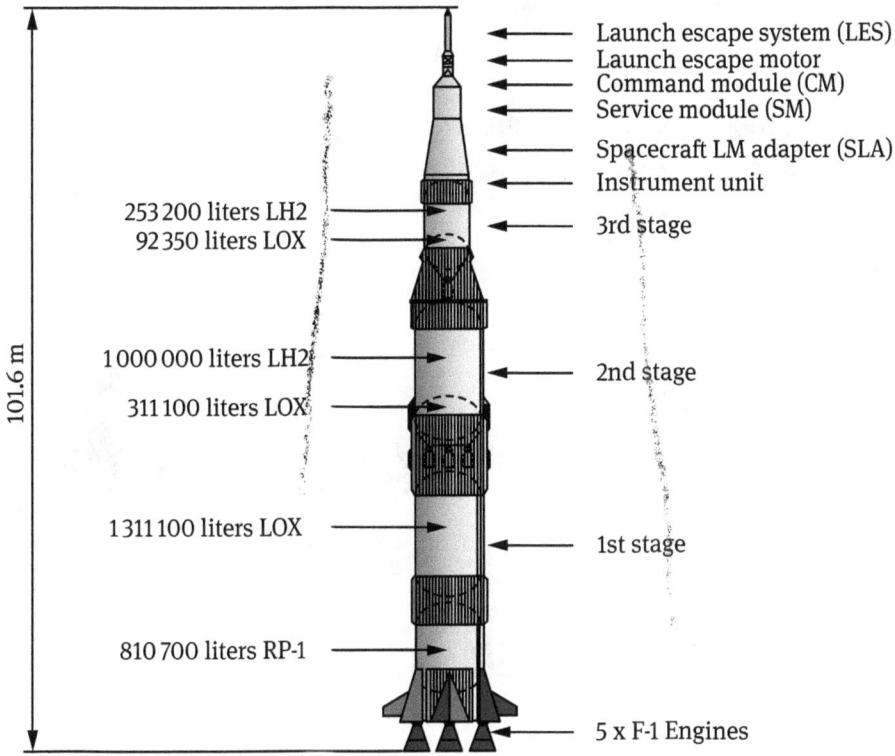

Figure 9.1: Stages and fuel masses of the Apollo Saturn V lunar rocket.

The Saturn V rocket was a three-stage liquid-fueled launch vehicle which was developed to support the Apollo programme for exploration of the Moon. The Apollo programme was the third United States human space flight program carried out by the National Aeronautics and Space Administration (NASA). Later the vehicle was used to launch the American Skylab space station. The Saturn V launch vehicle was launched 13 times from Kennedy Space Center (KSC) with no losses of crew or payloads. Only the final test flight of an unmanned Apollo 6 Saturn V rocket did not run smoothly because the rocket experienced severe pogo oscillations.

The term 'pogo' is a reference to the bouncing of a pogo stick, which is a device for jumping off the ground in a standing position through the aid of a spring. Pogo oscillations in liquid fuel rocket engines arise from thrust fluctuations in the engines. These oscillations of the engine performance occur during flight, as a result of pressure changes in the engine, and cause variations of acceleration on the structure of the rocket, giving variations in fuel pressure and flow rate. If this oscillation meet the resonant frequency of the overall system or the propellant system, the oscillations may increase to an extent that the vehicle structure is self-destructed.

Figure 9.2: Apollo 11 Saturn V lift-off on July 16, 1969. Credit: NASA [10].

9.1.2 Ariane 5 launch vehicle

After expiring of the Ariane 4 rocket, the European launcher Ariane 5 shall be marketed as commercially operating launcher system. The operating Arianespace company uses the CSG spaceport in Kourou, French-Guiana to launch satellites and payloads. Table 9.2 gives some characteristics of the Ariane 5 launch vehicle.

Table 9.2: Characteristics of the Ariane 5 launch vehicle.

Characteristic	Main stage (EPC)	Booster (EAP) each	Upper stage (ESC-A)
Mass, complete	170.3 t	278.33 t	19.44 t
Mass, propellant	158.1 t	237.5 t	14.9 t
Mass, dry	12.2 t	38.2 t	4.54 t
Propellant(s)	LOX / LH2	HTPB	LOX / LH2
Engine	Vulcain 2	P241	HM7-B
Thrust	960 kN (sea level)	7 080 kN (vacuum)	67 kN
	1390 kN (vacuum)		
Firing time	540 s	130 s	945 s

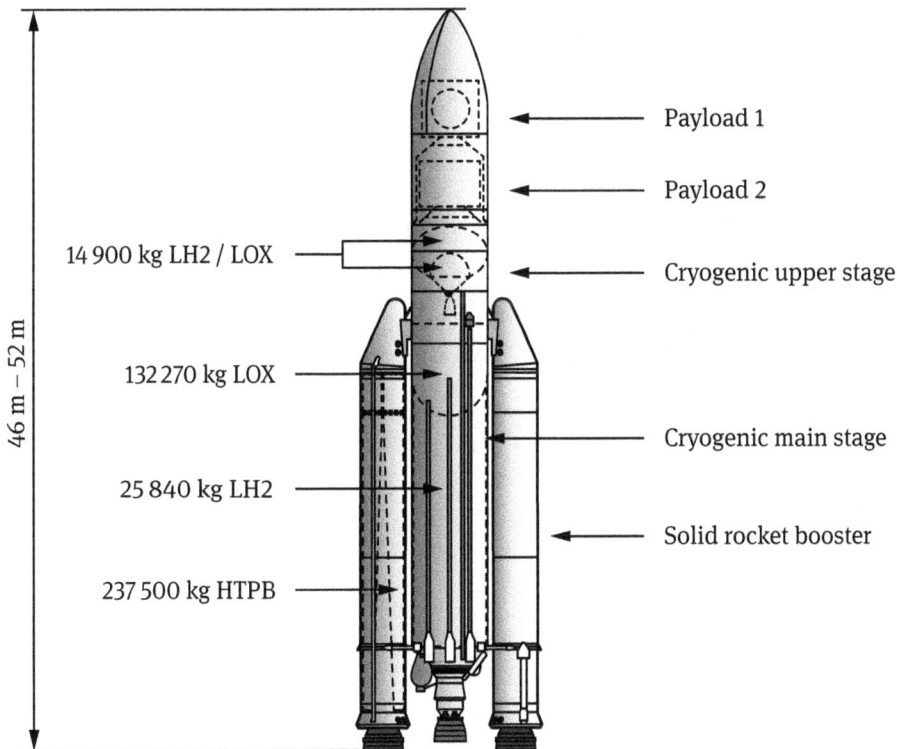

Figure 9.3: Stages and fuel masses of the European Ariane 5 launch vehicle.

Figure 9.4: Ariane 5 lift-off on flight VA225 on 20 August 2015. Image: ©ESA/CNES/ARIANESPACE-Photo Optique Video CSG.

Meanwhile, more than seventy Ariane 5 launchers ensured a guaranteed and sustainable European access to space. Figure 9.3 shows the design of the Ariane 5 launcher system. At present there are two cofigurations, the Ariane 5 ECA (Evolution Cryotechnique type A) and the Ariane 5 ES (Evolution Storable), which is used to deliver payloads into geostationary transfer orbit (GTO) or low Earth orbit (LEO).

The Ariane 5 ECA is currently the only commercial heavy-lift launcher which is capable to deliver two telecommunication satellites as payloads into GTO at the same time. The Ariane 5 ES launcher is used for different missions. Deploying a cluster of four operational Galileo satellites of the global navigation satellite system (GNSS) into medium Earth orbit (MEO) or placing ESA's Automated Transfer Vehicle into LEO for re-supply of the International Space Station. The ATV is injected into a 260 km circular low orbit with an inclination of 56° to the Equator. After reaching the orbit the ATV uses its own engine to reach the International Space Station.

The main differences of the Ariane 5 ES are a restartable storable propellant upper stage, which can perform multiple thrusts to place payloads into the desired orbit, and a reinforced vehicle equipment bay (VEB) to withstand flight loads with the ATV.

9.1.3 Ariane 6 Launch Vehicle

As of 2015, the new-generation launcher Ariane 6 is being developed by ESA and the European industry. The first flight is scheduled in 2020 and a fully operational level is set for 2023. Depending on the propulsion of the main stage and the number of additional boosters (two or four) the launcher covers a wide range of missions.

- Injecting satellites directly into GEO or via intermediate GTO and LEO.
- Injecting satellites into polar orbit or SSO.
- Injecting satellites into MEO or setting a spacecraft into Mars transfer orbit (MTO) on a trajectory to planet Mars.

The concept of the Ariane 6 launcher provides for two PHH configurations. The first stage uses either two (Ariane 62) or four (Ariane 64) P120 solid-propellant boosters depending on the requested performance. The boosters of the first stage are based on solid propulsion (P) and the second and third stage use cryogenic LOX/LH2 propulsion (H). The total length of the Ariane 6 launch vehicle is approximately 63 m and the cryogenic main stage is loaded with approximately 149 t of propellants. The external diameter of the main stage including upper stages and fairing connections is approximately 5.4 m.

On 12 August 2015 Airbus Safran Launchers as prime contractor for Ariane 6 and the European Space Agency (ESA) signed a € 2.4 billion (in total € 3 billion) contract to develop the Ariane 6 launcher in two versions, Ariane 62 and Ariane 64.

Table 9.3: Characteristics of the Ariane 6 launch vehicle (subject to change).

| Characteristic | Main stage | | Booster (each) | Upper stage |
	A62	A64		A62 / A64
Mass, complete	~500 t	~800 t	n.s.	19.44 t
Mass, propellant	149 t	149 t	120 t	30 t
Mass, empty	~360 t	~650 t	n.s.	n.s.
Propellants	LOX / LH2	LOX / LH2	NH_4ClO_4 / Al, HTPB	LOX / LH2
Engine	Vulcain 2.1	Vulcain 2.1	P120C	Vinci (re-ignitable)
Thrust	1350 kN[12]	1350 kN[12]	3500 kN[12]	180 kN[13]
Firing time	~540 s	~540 s	130 s	~900 s

The two Ariane 6 variants planned are shown in Figure 9.5 below. The Ariane 6 launcher is so heavy that at least two solid rocket boosters are necessary to allow the rocket to lift off.

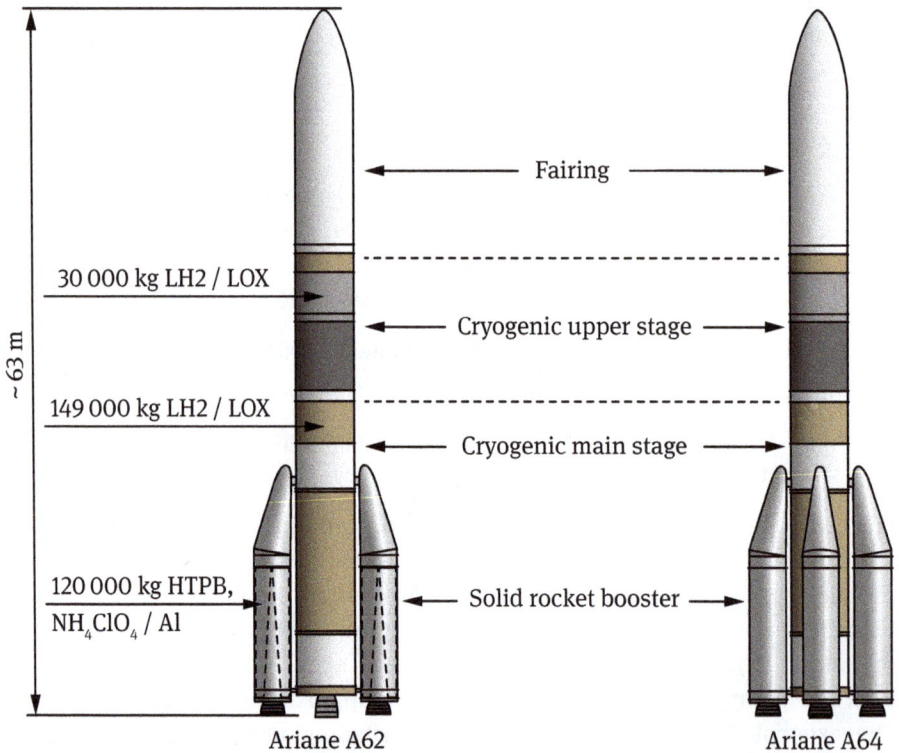

Figure 9.5: Stages and fuel masses of the European Ariane 6 launch vehicle (subject to change).

---

12 At sea level.
13 In the vacuum.

Figure 9.6: Artist's impression of the Ariane A64 configuration. Image: ©ESA – David Ducros [11].

9.1.4 Space Transportation System

The Space Shuttle programme, officially called Space Transportation System (STS), was a manned launch vehicle programme of the United States government from 1981 to 2011. A Space Shuttle orbiter was launched with two re-usable solid rocket boosters (SRBs) and a disposable external fuel tank. Usually, the orbiter carried four to seven astronauts and up to 22.7 t of payload into low Earth orbit (LEO).

Table 9.4: Characteristics of the Space Transportation System.

Characteristic	External tank	Booster (SRB)	Orbiter
Mass, complete	765 t	590 t	19.44 t
Mass, propellant	735.6 t	498.95 t per SRB	14.9 t
Mass, empty	26.5 t	83.92 t per SRB	74.84 t
Propellant(s)	LOX / LH2	PBAN,-APCPB	LOX / LH2
Engine	–	–	3 x RS-25 (Rocketdyne)
Thrust (per SRB)	–	12500 kN[14]	1860 kN^2 / 2279 kN[15]
Firing time	–	124 s	480 s

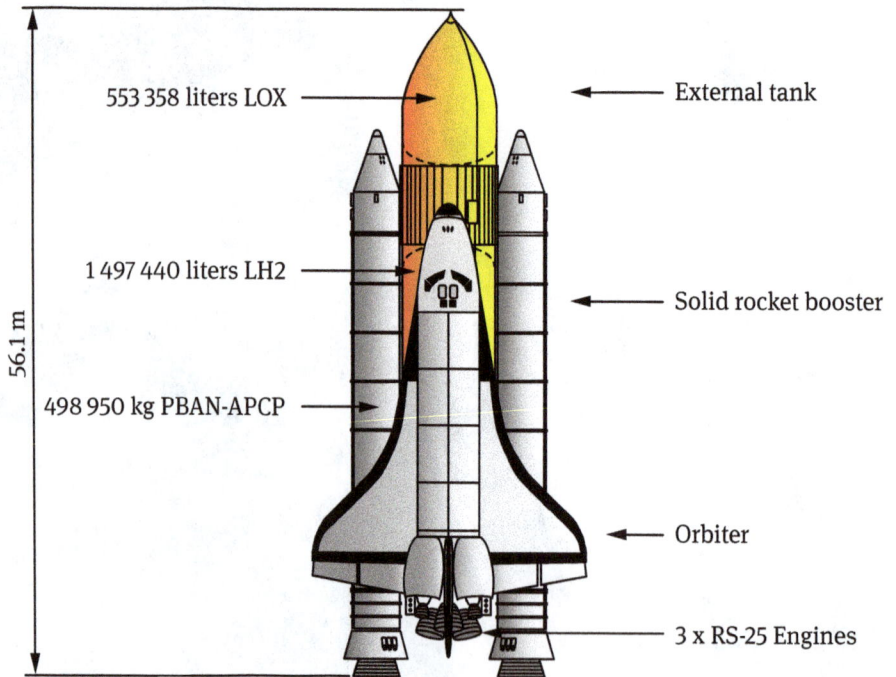

553 358 liters LOX ⟶ ◀—— External tank

1 497 440 liters LH2 ⟶

56.1 m

498 950 kg PBAN-APCP ⟶ ◀—— Solid rocket booster

◀—— Orbiter

—— 3 x RS-25 Engines

Figure 9.7: American Space Transportation System (STS).

---

14 At sea level.
15 Vacuum.

Figure 9.8: Space Shuttle *Enterprise* atop NASA's Shuttle Carrier Aircraft (SCA). Credit: NASA [12].

The Space Shutte programme was administered by the National Aeronautics and Space Administration (NASA) responsible for the civilian space programme as well as for scientific research in aeronautics and aerospace. It was given responsibility for construction and supply of the International Space Station (ISS). The missions involved also the deployment, retrieval, and repair of satellites.

The first Space Shuttle prototype *Enterprise* was an experimental orbiter, launched from the back of a modified Boeing 747 aircraft to perform approach and landing tests to verify aerodynamic and control characteristics, especially in preparation for the first mission of the *Columbia* orbiter in space. Space Shuttle *Enterprise* taken aloft on NASA's Boeing 747 Shuttle Carrier Aircraft (SCA) is shown in the above Figure 9.8.

After the destruction of the *Columbia* and the explosion of the *Challenger* three remaining space shuttles were used for 135 manned space missions.

Table 9.5: Space Shuttle orbiters.

Name	First flight	Status
Enterprise	As of 1977	Only for attempted landings / ground tests
Columbia	12 December 1981	Lost on 1 February 2003 during landing
Challenger	4 April 1983	Lost on 28 January 1986 during lift-off
Discovery	30 August 1984	In service until 2011
Atlantis	3 October 1985	In service until 2011
Endeavour	7 May 1992	In service until 2011

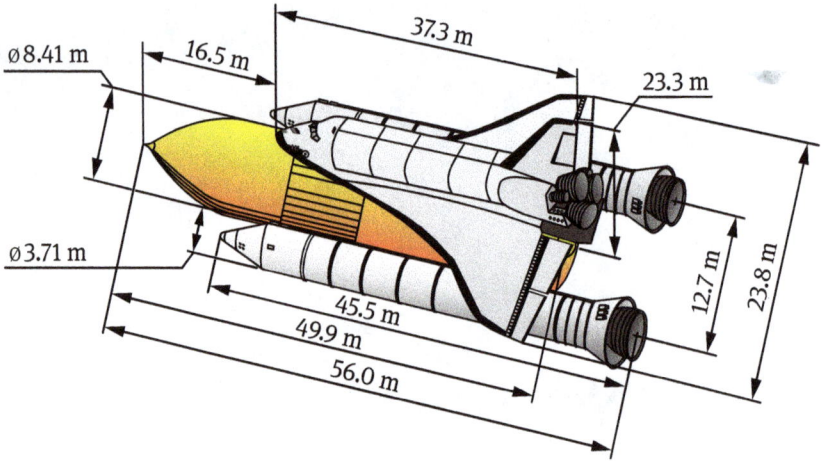

Figure 9.9: Dimensions of the Space Transportation System (STS).

The Space Shuttle *Endeavor* had an operational altitude of 190 km to 960 km. The shuttle's velocity was 7743 m·s⁻¹. The size of the Space Shuttle *Endeavor* gives it a maximum payload of 25060 kg. The payload bay specifications are 4.6 m x 18 m. The dimensions if the orbiter, boosters and external tank are shown in Figure 9.9.

The reaction control system (RCS) is used for attitude control. Attitude refers to the orientation of the orbiter on all three spatial axes. The RCS engines control the attitude of the orbiter by affecting rotation around the three axes. Rotating the nose clockwise or counterclockwise is referred to as 'roll', moving the nose up and down is referred to as 'pitch', and moving the nose left and right is referred to as 'yaw' (see Figure 9.10 below).

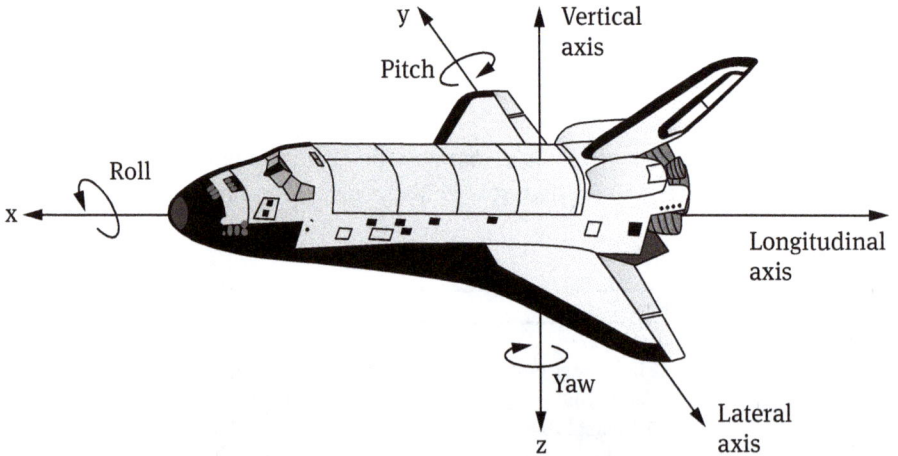

Figure 9.10: Diagram of the x-, y-, and z-axes of the Space Shuttle.

Figure 9.11: Space Shuttle *Endeavor*, STS-99 lift-off on 11 February 2000. Credit: NASA [13].

A large number of thrusters for manoeuvering of the Space Shuttle were combined in an attitude control system or reaction control system (RCS) and a propulsion system for orbit changes or orbital manoeuvering system (OMS).

- 38 x 3 870 N thrusters and 6 x 110 N thrusters (RCS)
- 2 x 26 700 N thrusters (OMS)

The thrusters of the reaction control system (RCS) are small rocket engines allowing the orbiter to change attitude, and to perform various manoeuvres in the orbit such as movements around the rotational axes (Figure 9.12) and around the translational axes, left/right, forward/backward and up/down.

The RCS of the orbiter consists of high-pressure gaseous helium storage tanks, pressure regulation and relief systems, and a fuel and oxidiser tank. The locations of the forward RCS thrusters and the RCS/OMS thrusters of the aft fuselage for manoeuvering of the Shuttle are illustrated in the Figures 9.12 and 9.13, respectively. An additional system distributes the propellant to the engines and a thermal control system (electrical heaters). The heaters prevent the propellant from freezing in the tanks and lines. Each heater system has two redundant heater systems serving as failover systems.

The two helium tanks in each RCS supply gaseous helium to the fuel tank and to the oxidiser tank, respectively. The helium expelled the propellant into a propellant acquisition device. The propellant acqusition device delivers the propellant to the RCS. The forward RCS propellant tanks have propellant acquisition devices designed to operate under microgravity as in low Earth orbit. Whereas the aft RCS propellant tanks are designed to operate in both 1g-gravity and microgravity to ensure propellant and pressurant separation during tank operation.

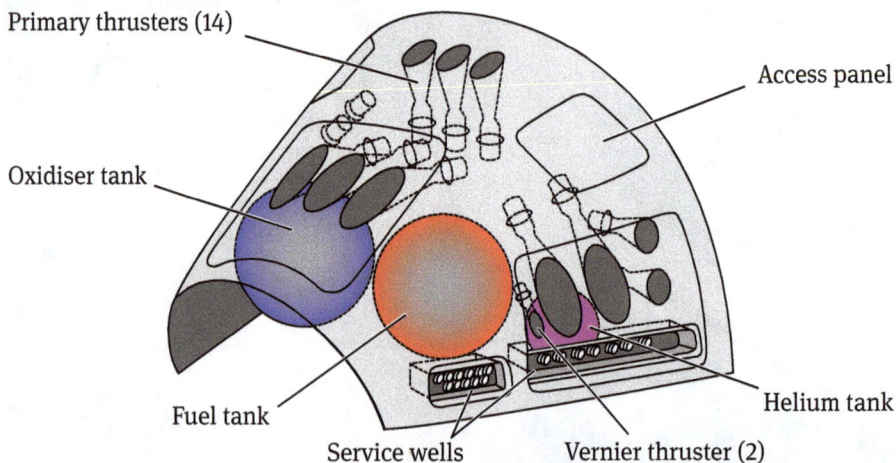

Figure 9.12: Forward RCS thruster locations of the Space Shuttle orbiter.

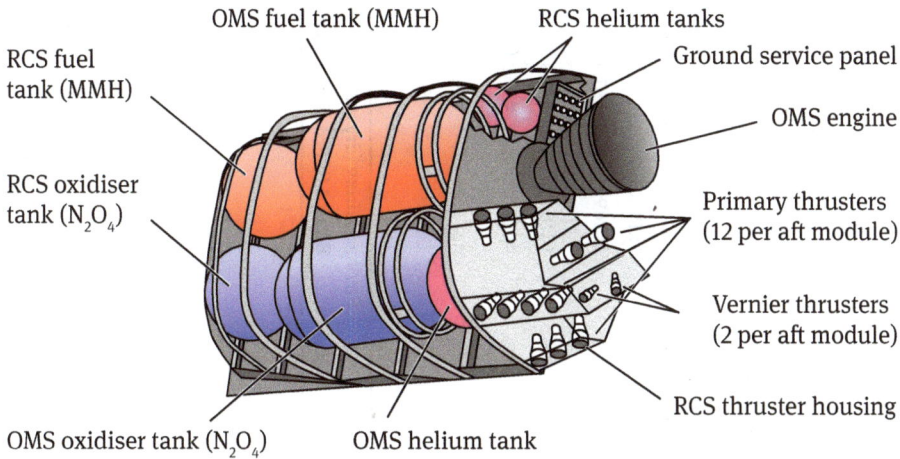

Figure 9.13: RCS/OMS thrusters of the left aft fuselage of the Space Shuttle orbiter.

The orbital manoeuvring system (OMS) provides thrust for orbit insertion, orbit circularisation, orbit transfer, rendezvous and de-orbiting. The OMS is housed in the left and right aft fuselage pod of the orbiter (Figure 9.13). Both the OMS engines and the RCS used a hypergolic propellant combination of monomethylhydrazine (MMH) as fuel and dinitrogen tetroxide (N_2O_4) as oxidiser.

9.1.5 Space Launch System

NASA's Space Launch System (SLS) will be the most powerful rocket to transport up to four astronauts in the Orion Spacecraft deeper into the Solar System than ever before. One of the big plans of NASA is to send astronauts to an asteroid by the year 2025. This asteroid mission will be performed by NASA's Space Launch System and the Orion spacecraft to fly astronauts beyond low-Earth orbit. The Orion spacecraft will not only be used as an exploration vehicle, but also for sustaining the crew during the mission and providing a safe return from space.

On 4 September 2011, NASA announced the design of the Space Launch System with three planned versions (Figure 9.14). All three versions use the same core stage with four RS-25D/E main engines. Block 1B will feature a more powerful exploration upper stage (EUS) than the interim cryogenic propulsion stage (ICPS) of Block 1 as second stage. The Blocks will use two five-segment solid rocket boosters (SRBs) operating in parallel with the main engines for the first 120 s of flight to provide an additional thrust needed for escape from the gravitational pull of the Earth.

The LEO payload capacity of Block 1 will be 70 t, and Block 1B will has a payload capacity of 105 t. The Block 2 will have a lift capacity of 130 t due to advanced boosters and a different Earth departure stage (EDS) with up to three J-2X engines.

Table 9.6: Characteristics of NASA's Space Launch System (subject to change).

Characteristic	Five-segment booster (each)			Core stage (First stage)			Upper stage (Second stage)		
	Block 1	Block 1 (ICPS)	Block 2	Block 1	Block 1 (ICPS)	Block 2	Block 1	Block 1 (ICPS)	Block 2
Mass, complete	729.8 t	731.9 t	~793 t	1068.3 t	1091.45 t	1091.54 t	n.s.	n.s.	n.s.
Mass, propellant	626.1 t	631.5 t	~709 t	978.9 t	979.45 t	966.071 t	n.s.	n.s.	n.s
Mass, dry	103.7 t	100.4 t	~84 t	76.1 t	~102 t	115.577 t	n.s.	3.7649 t	26.40 t
Propellant(s)	PBAN – APCP	PBAN – APCP	HTPB	LOX / LH2	LOX / LH2	LOX / LH2	LOX / LH2	LOX / LH2	LOX / LH2
Engine	n/a	n/a	n/a	4 × RS25D	4 × RS25D	4 × RS25E	n.s.	1 × RL10B-2	2 × J2X
Thrust	1592.5 t	1428.8 t	2041.0 t	758.4 t (s. l.) 929.6 t (vac.)	758.4 t (s. l.) 929.6 t (vac.)	758.4 t (s. l.) 938.04 t (vac.)	n.s.	112492 t	261.27 t
Firing time	126.6 s	128.4 s	110 s	476 s	476 s	476 s	n.s.	1118 s	344 s
Specific impulse	237 s (s. l.) 267.4 s (vac.)	237 s (s. l.) 267.4 s (vac.)	~259 s (s. l.) 286 s (vac.)	366 s (s. l.) 452.1 s (vac.)	366 s (s. l.) 452.1 s (vac.)	366 s (s. l.) 452.4 s (vac.)	n.s.	461.5 s (vac.)	436 s (vac.)
Diameter	3.71 m	3.71 m	3.71 m	8.38 m	8.38 m	8.38 m	n.s.	n.s.	n.s.
Height	53.87 m	53.87 m	53.87 m	62.54 m	62.54 m	62.54 m	n.s.	13.7 m	~23 m

Characteristic	SLS (complete)			Interstage			Payload fairing (dry mass)		
	Block 1	Block 1 (ICPS)	Block 1A / 2	Block 1	Block 1 (ICPS)	Block 1A / 2	Block 1	Block 1 (ICPS)	Block 1A / 2
Mass	2650 t	2650 t	2700 t / 2950 t	n/a	4.99 t	5.22 t	~10.6 t	8.17 t	12.11 t
Height (incl. payload)	92.3 m	97.56 m	~113 m						
Height (w/o payload)	64.7 m	64.7 m	~87 m						
Payload to 296 km	>70 t (~95 t)		~105 t / 145 t						
Payload to TLI		24.5 t	~45 t / 60 t						

s. l. = sea level
vac. = vacuum
n.s. = not specified
n/a = not applicable
TLI = trans lunar injection

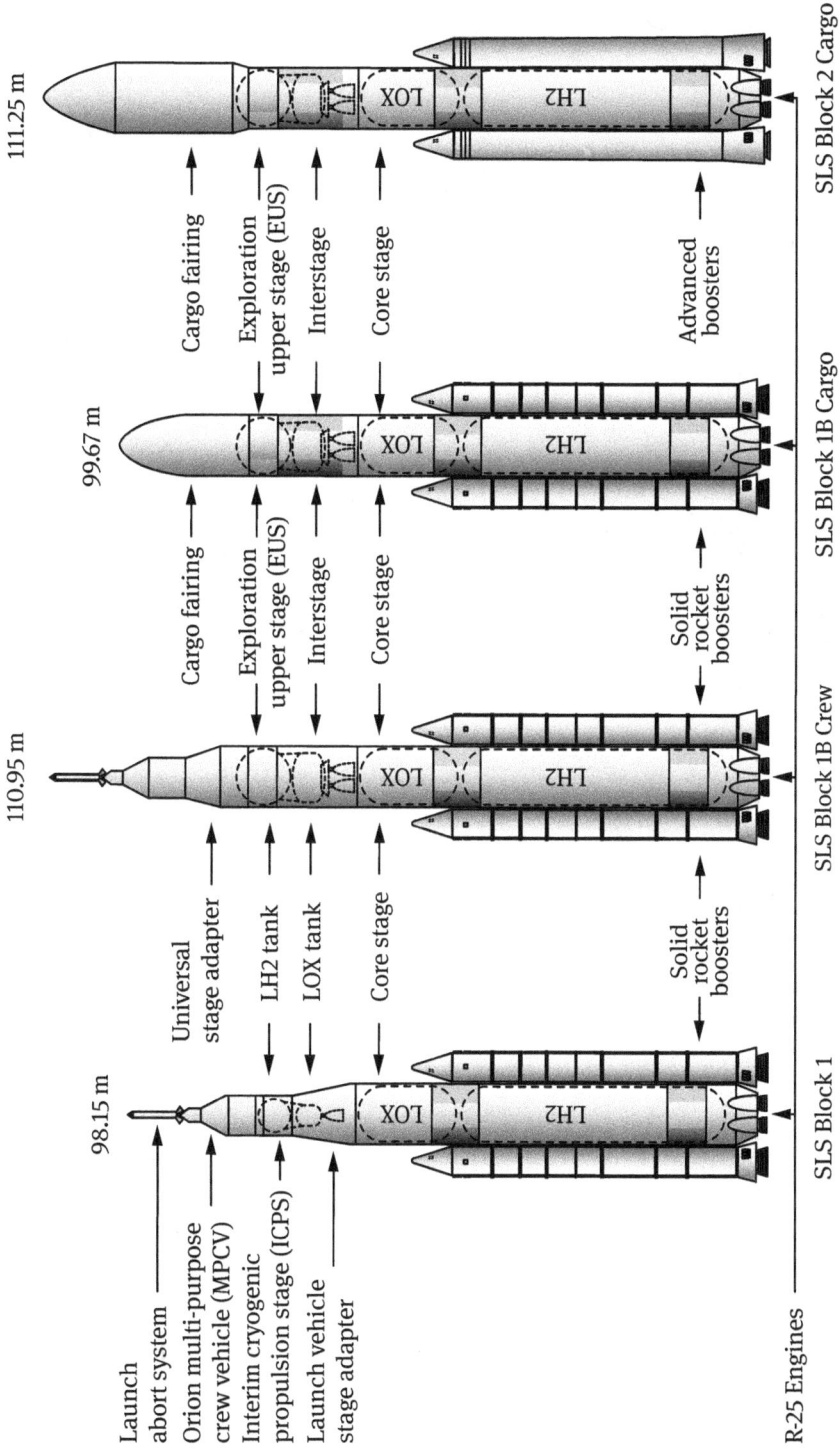

Figure 9.14: Configuration of the Space Launch System (SLS).

Figure 9.15: Artist's view of SLS Block 1 (configuration at critical design review). Credit: NASA [14].

9.1.6 Other launcher systems

A number of launcher systems from past to present are shown in Table 9.7.

Table 9.7: Launcher systems from different countries (not complete).

Name	Country	First flight	Remark
Soyuz	Soviet Union	4 October 1957	Sputnik 1
Redstone	United States	31 Januar1958	Explorer 1, decommissioned
Vostok	Soviet Union	2 January1959	Decommissioned
Atlas	United States	9 September 1959	Boeing
Delta	United States	13 May 1960	Boeing
Scout	United States	1 June 1960	Decommissioned
Molniya	Soviet Union	19 February 1964	Luna 9 to soft-land on the Moon
Titan	United States	8 April 1964	Decommissioned
Diamant	France	26 November 1965	Asterix 1, decommissioned
Tsyklon	Soviet Union	17 September 1966	Decommissioned in 1969.
Cosmos	Soviet Union	15 May 1967	Last launch in 2010
Saturn V	United States	11 September 1967	Decommissioned
Proton	Soviet Union	16 November 1968	Still in use as of 2015
Lambda	Japan	2 November 1970	Japan's first satellite launcher
Long March	China	24 Apri 1970	Expendable launch system
ELDO	Europe	12 July 1970	From Woomera, decommissioned
Ariane-1	Europe	24 December 1979	Arianespace
SLV	Indien	20 September 1980	Retired satellite launch vehicle
Space Shuttle	United States	12 April 1981	Decommissioned
Zenit	Soviet Union	13 April 1985	Also for sea-launch
H-II	Japan	3 February 1994	Used domestical technologies
ASLV	India	24 March 1987	Decommissioned
Shavit	Israel	19 September 1988	From Palmachim
Energija/Buran	Soviet Union	15 November 1988	Decommissioned
Pegasus	United States	5 April 1990	Air-Launch
Start	Russia	25 March 1993	Launch from mobile platform
PSLV	India	20 September 1993	For polar orbits
Taurus	United States	13 March 1994	Lockheed Martin
Rockot	Russia	26 December 1994	SS-19
Athena	United States	1 July 1995	Lockheed Martin
Shtil	Russia	7 July 1998	SS-N-23 from submarine
Taepodong	North Korea	31 August 1998	From No-Dong
Dnepr	Russia	21 April 1999	Converted ICBM
GSLV	India	18 April 2001	For geostationary orbits
Volna	Russia	12 July 2002	SS-N-18 from submarine
Safir	Iran	3 February 2009	First expendable launch vehivle
KSLV	South Korea	25 August 2009	Failed to reach orbit
Falcon 9	United States	4 June 2010	Manufactured by SpaceX
Vega	Europe	13 February 2012	Geodatic and nanosatellites
Antares	United States	21 April 2013	Cygnus mass simulator
Angara	Russia	9 July 2014	Mass simulator
VLS-1 V-04	Brazil	2018	Planned from Alcântara

9.2 Satellites and Probes

Since the beginning of space flight in 1957, several thousand satellites have been launched. These satellites can be rougly divided according to

- orbit and application,
- manned or unmanned, and others.

For a larger list of satellites launched, refer to Chapter 10 (Projects and Payloads). Exemplarily, the Hubble Space Telescope and the Cassini space probe will be presented here as representatives for unmanned satellites and the Apollo 11 service and command module as a manned satellite.

9.2.1 Hubble Space Telescope

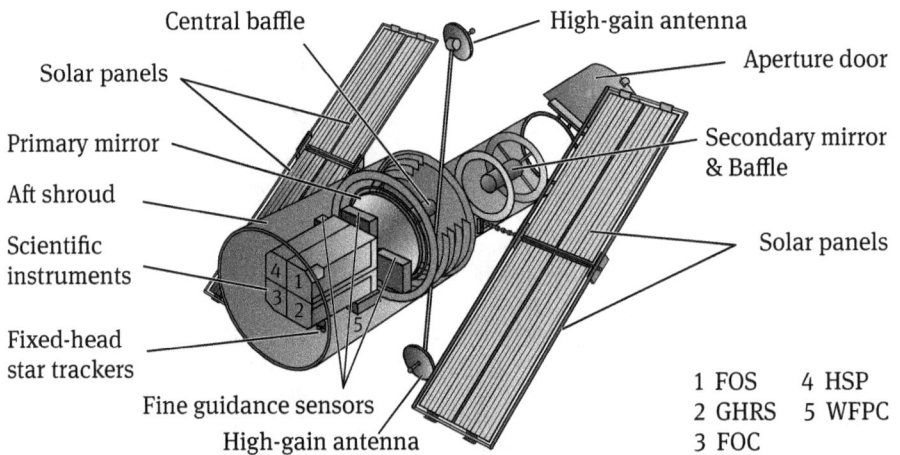

Figure 9.16: Initial instruments of the Hubble Space Telescope.

On 26 April 1990 the Hubble Space Telescope was deployed into low Earth orbit from the payload bay of the Space Shuttle *Discovery* (STS-31) and was reguarly maintained, as required, by means of manned flights of the American Space Transportation System. When launched, the Hubble Space Telescope carried five scientific instruments for observations. The Faint Object Spectrograph (FOS), the Goddard High Resolution Spectrograph (GHRS), the Faint Object Camera (FOC), the High Speed Photometer (HSP), and the Wide Field/Planetary Camera (WFPC). Each scientific instrument is located in different portions of the telescope's focal plane (Figure 9.16). After the Fixed-Head Star Trackers (FHST) made a coarse-pointing to the target, three Fine Guidance Sensors (FGS) centre to the target observed.

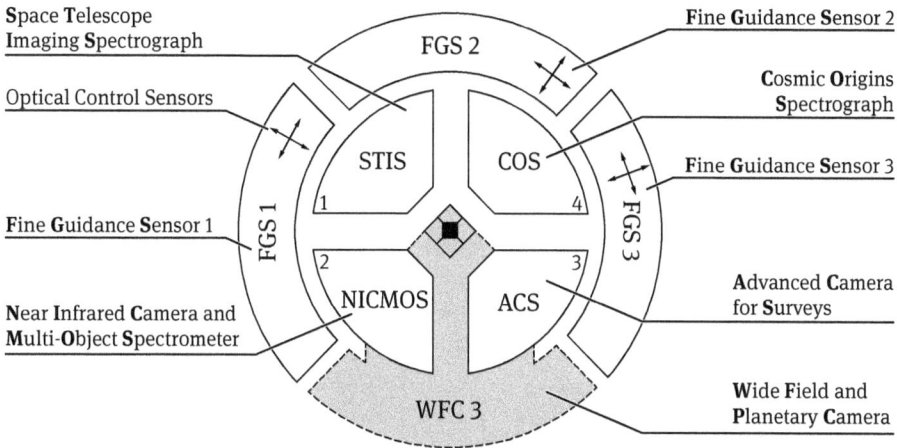

Figure 9.17: Instrument apertures of the telescope's focal plane after the last Servicing Mission 4 in May 2009.

The light path within the telescope is illustrated in Figure 9.17. The concave primary mirror of the telescope measures 2.4 m in diametre and weighs approximately 828 kg. It captures light from objects in space and focuses it towards the secondary mirror. The convex seconday mirror measures 0.3 m in diametre and weighs approximately 12.3 kg. It redirects the light coming from the primary mirror through a hole in the centre of the primary mirror onwards to the scientific instruments.

The scientific instruments of the Hubble Space Telescope are designed to record and measure light from infrared through visible into ultraviolet wavelengths, simultaneously. The telescope does not magnify objects but actually collecting more

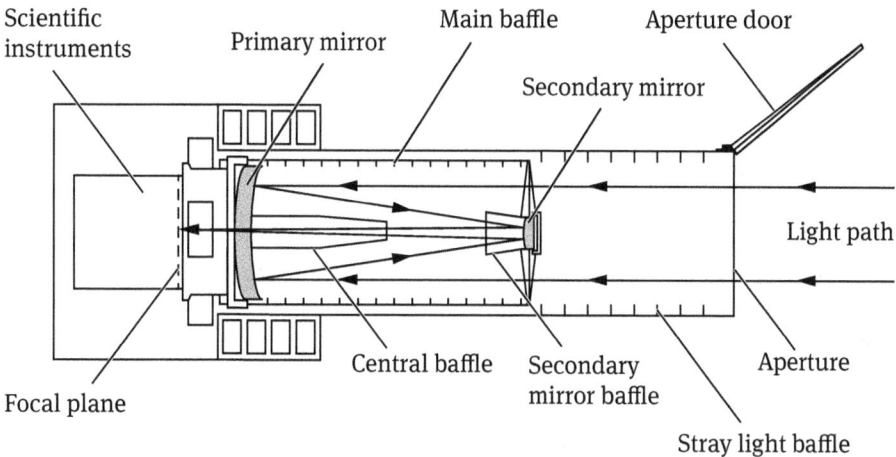

Figure 9.18: Light path within the Hubble Space Telescope.

light than the eyes of a human can capture. As of 2015, the Hubble Space Telescope is still operating and may last until 2030 to 2040. The successor of the Hubble Space Telescope will be the James Webb Space Telescope as the 'Next Generation Space Telescope' which is scheduled for launch in October 2018.

Corrective Optics Space Telescope Replacement (COSTAR)

The Corrective Optics Space Telescope Axial Replacement, installed during the 1993 Servicing Mission 1 (mission STS-61) was not an instrument but provided correcting optics for all three originally installed scientific instruments. The HSP instrument was removed and made space for COSTAR. Two of the instruments were replaced by STIS and NICMOS which have built-in optical corrections. COSTAR corrected images only for the FOC instrument and was no longer be needed when the FOC instrument was replaced. During Servicing Mission 4 (mission STS-125) in 2009, astronauts removed COSTAR and had been replaced by the Cosmic Origins Spectrograph (COS).

Space Telescope Imaging Spectrograph (STIS)

The Space Telescope Imaging Spectrograph (STIS) breaks light into its wavelengths and determines the chemical composition, temperature, speed, and density of objects. The spectrograph records light in multiple wavelengths with high resolution in the ultraviolet wavelengths from hundreds of different positions, simultaneously. The task of the spectrograph is the mapping of planets, galaxies, and nebulae, searching for black holes and examining quasars and comets. The STIS instrument was installed during Servicing Mission 3B (mission STS-109) in March 2002 and replaced the FOS instrument.

Cosmic Origins Spectrograph (COS)

The Cosmic Origins Spectrograph is a sensitive ultraviolet spectrograph and focuses exclusively on ultraviolet light. The telescope's sensitivity was improved at least 10 times in ultraviolet light and up to 70 times when the telescope is looking at extremely faint objects. Astronauts of Space Shuttle *Atlantis* replaced COSTAR by COS instrument during the final Servicing Mission 4 (mission STS-125) in May 2009.

Near-Infrared Camera and Multi-Object Spectrometer (NICMOS)

The Near-Infrared Camera and Multi-Object Spectrometer is an advanced infrared detector which allows to see through interstellar dust, to view stars and planet formation. NICMOS also allows studies of faint galaxies and was installed during Servicing Mission 2 (STS-82) in February 1997. The three cameras of the NICMOS instrument operate independently. The NICMOS instrument also operates as a spectrometer, coronograph, and polarimeter and searches for planets in other solar systems and observes star birth.

Advanced Camera for Surveys (ACS)

The Advanced Camera for Surveys is the newest digital camera installed during Servicing Mission 3B (mission STS-109) in March 2002, and replaced the FOC instrument. The ACS instrument has a wider field of view and better light sensitivity than the WFPC2 instrument. It effectively increases telescope's discovery power by 10 times. The wide field camera in the ACS instrument is a 16 megapixel camera.

Wide Field/Planetary Camera 3 (WFPC3)

The Wide Field/Planetary Camera 3, installed as replacement of WFPC2 during Servicing Mission 4 (mission STS-125) in March 2009, consists of four cameras with relay mirrors to correct the flaw in the primary mirror. The tasks of the WFPC3 are the search for new planets, imaging the neighbouring planets, refining distances to other galaxies, and providing data to determine the age of the universe.

9.2.2 Cassini Space Probe

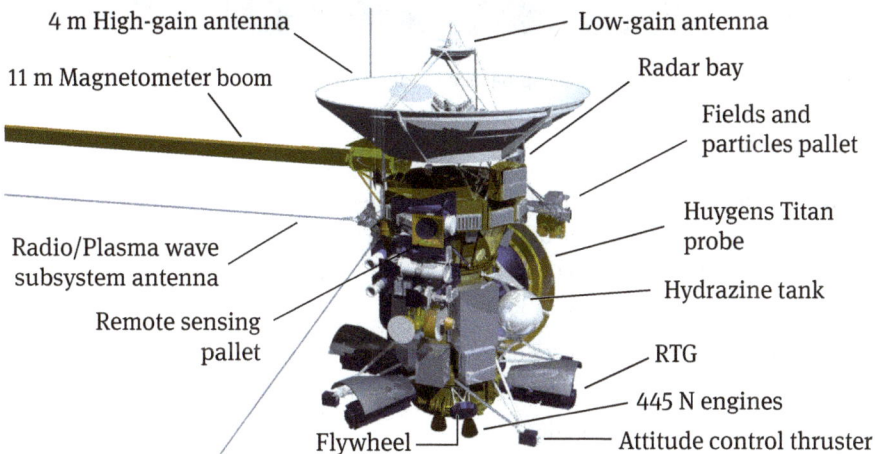

Figure 9.19: Assemblies of Cassini spacecraft with ESA's Huygens Titan lander. Credit: NASA [15].

Launched in 15 October 1997, the Cassini spacecraft is the largest space probe sent into space than ever before. The Cassini mission is an endeavor of NASA, the European Space Agency (ESA) and the Agenzia Spatiale Italiana (ASI) to study Saturn's atmosphere, magnetosphere, moons, and its ring system. The Cassini spacecraft carried ESA's Huygens Titan lander to study *Titan*, Saturn's largest moon. The Huygens probe landed on Titan's surface on 14 January 2005 after collecting data of Saturn's moon *Phoebe* while Cassini's instruments continuously returning data from Saturn's system since the arrival at Saturn in July 2004.

Figure 9.20: Artist's impression of the approach of Cassini spacecraft at Saturn. Credit: NASA [16].

9.2.3 Apollo Modules

The Apollo programme is exemplified for manned satellites. For Moon flights an appropriate powerful launcher system, the Saturn V rocket, was developed. The payload of the rocket consisted of

– a service module (SM) with
– a command module (CM) as re-entry body for three astronauts,
– a two-staged lunar lander module (LM) for two astronauts consisting of
– a descent and an ascent module.

The essential subassemblies of the Apollo service and command module (CSM) and the descent and ascent stages are shown in Figure 9.21 to Figure 9.23. After launch, the 1rst and 2nd stage and also the launch escape system (LES) were jettisoned from the Saturn V launch vehicle. The remaining 3rd stage inserted the CSM into a lunar trajectory and separated itself from the rest of the vehicle. At that time, the spacecraft lunar module adapter (SLA) was jettisoned from the service module (SM). All that remained was the command module (CM), the service module (SM), and the lunar module (LM). The reaction control engines (RCEs) now adjusted its orientation so that the top of the CM faced a funnel-shaped device on the LM called the drogue.

Nose cone and 'Q-ball'

Pitch control motor

Canard assembly

Tower jettison motor

Launch escape motor

Launch escape tower

LAUNCH ESCAPE
ASSEMBLY

COMMAND
MODULE

SERVICE
MODULE

Environmental control
system radiator panel

Reaction control
thruster assembly

High gain antenna

Control nozzle

Service propulsion
engine nozzle

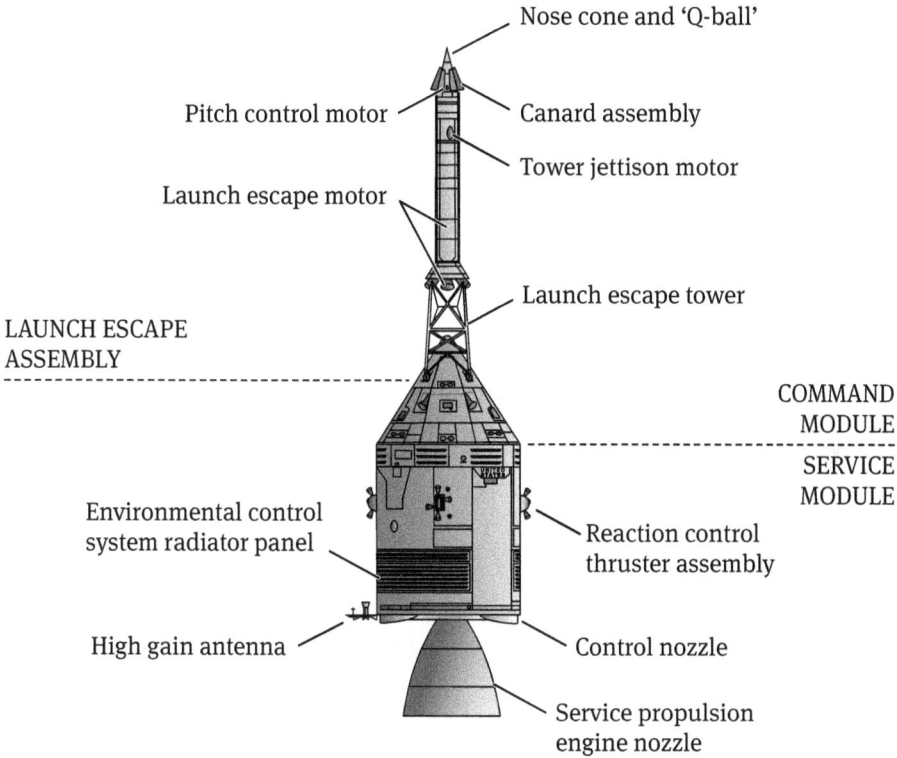

Figure 9.21: Apollo command and service module with launch escape assembly.

By means of a probe the CSM was aligned and docked with the LM's drogue. Once docked, the LM was secured with twelve automatic latches to the top of the CM. Then the probe and drogue assemblies were removed allowing the astronauts to move between the two modules.

The Apollo 11 lunar module *Eagle* was the first vehicle to land on the Moon. At first, the LM remained docked with the CSM during the voyage into the lunar orbit. Once in orbit, two of the three-astronauts moved from the CSM to the LM. Before undocking, both the LM and CM were sealed. The astronauts separated the two vehicles and the LM began its voyage to the surface of the Moon.

The lunar module consisted of two modules. The ascent stage was an irregularly shaped unit of approximately 2.8 m in height and 4.0 m by 4.3 m in width mounted atop of the descent stage with a constant-thrust engine of approximately 15 kN, and housed the astronauts in a volume of 6.65 m^3. Manoeuvring was achieved via the reaction control system (RCS), which consisted of four thruster assemblies, each one composed of four 450 N thrusters (Figure 9.22). The engine of the descent stage was a ablative cooled engine with a maximum thrust of approximately 45 kN mounted on a gimbal ring in the centre of the descent stage (Figure 9.23).

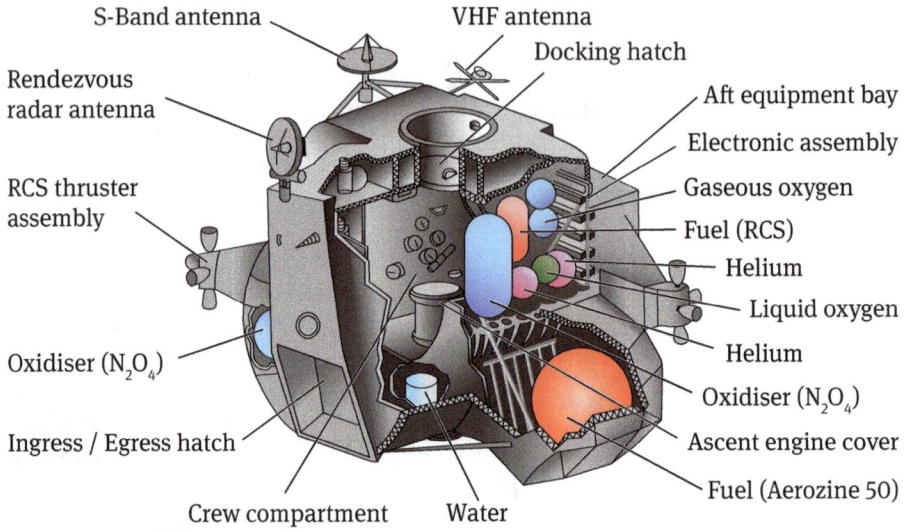

Figure 9.22: Apollo ascent module.

After landing on the Moon, the two astronauts prepared first the ascent stage for lift-off. After a resting time they prepared for the mission objectives on the Moon. After approximately 22 h they completed their mission objectives and returned to the LM for ascent. The ascent stage of the LM separated from the descent stage using explosive bolts. The reaction control system (RCS) of ascent stage provided enough thrust to launch it into the lunar orbit. The rendezvous radar antenna of the ascent stage received informations regarding the position and velocity of the CSM to allow docking.

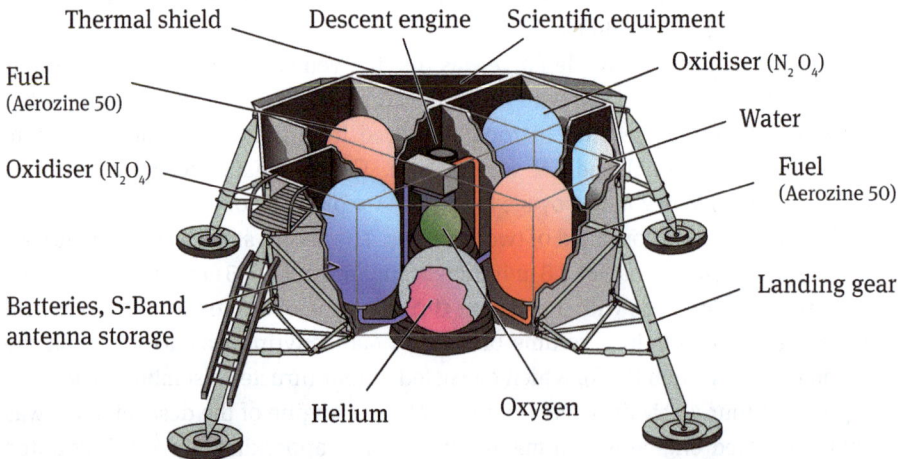

Figure 9.23: Apollo descent module.

9.3 Chemical Propulsion Systems

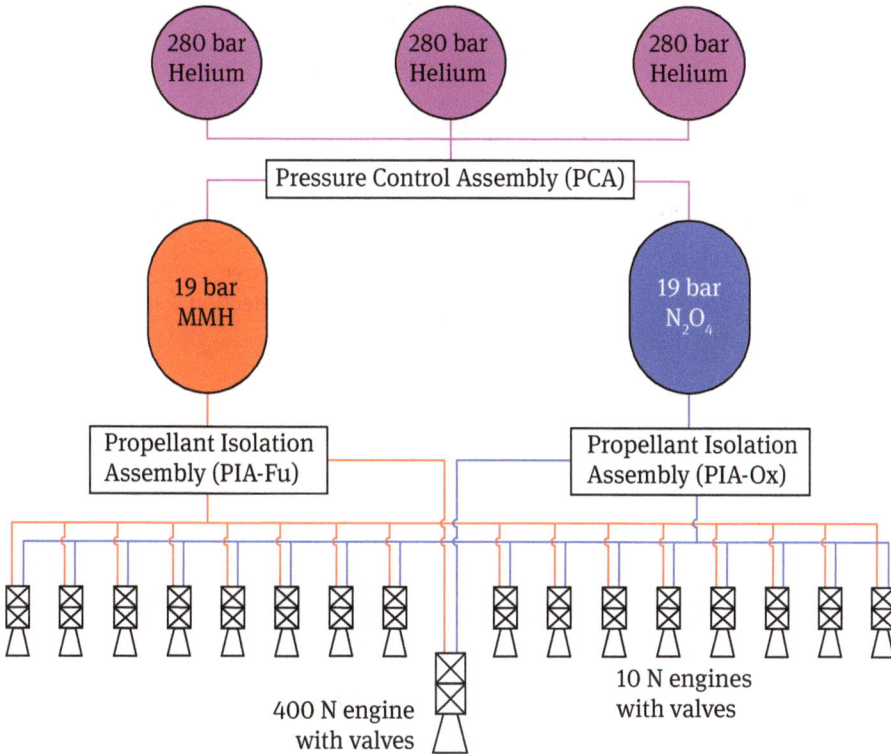

Figure 9.24: Basic structure of a bipropellant propulsion system.

From the abundance of the satellite's components, a simplified scheme of the bipropellant propulsion system of the European Alcatel Spacebus is exemplarily illustrated in the above Figure 9.24. This is a helium compressed gas supported propulsion system with storable propellants. The system consists of dinitrogen tetroxide (N_2O_4) as oxidiser and monomethylhydrazine (MMH, $N_2H_3 - CH_3$) as fuel and 14 to 16 10-N-thrusters for attitude control and a 400-N-thruster for injection of the satellite into orbit. This propulsion system is available with the name Unified Propulsion System (UPS) from Airbus DS.

Depending on the attitude and orbit control system of the satellite, simpler and thus more cost-effective monopropellant propulsion systems may be applied. The complete flow-sheet is shown in Figure 9.25. Also the Mars Reconnaissance Orbiter (see Chapter 13) used a monopropellant propulsion system for trajectory correction manoeuvres to keep the spacecraft on path to Mars and for orbit insertion manoeuvres to capture the spacecraft into orbit on arrival. Monopropellant propulsion systems are supplied commercially by Airbus DS for the satellite market.

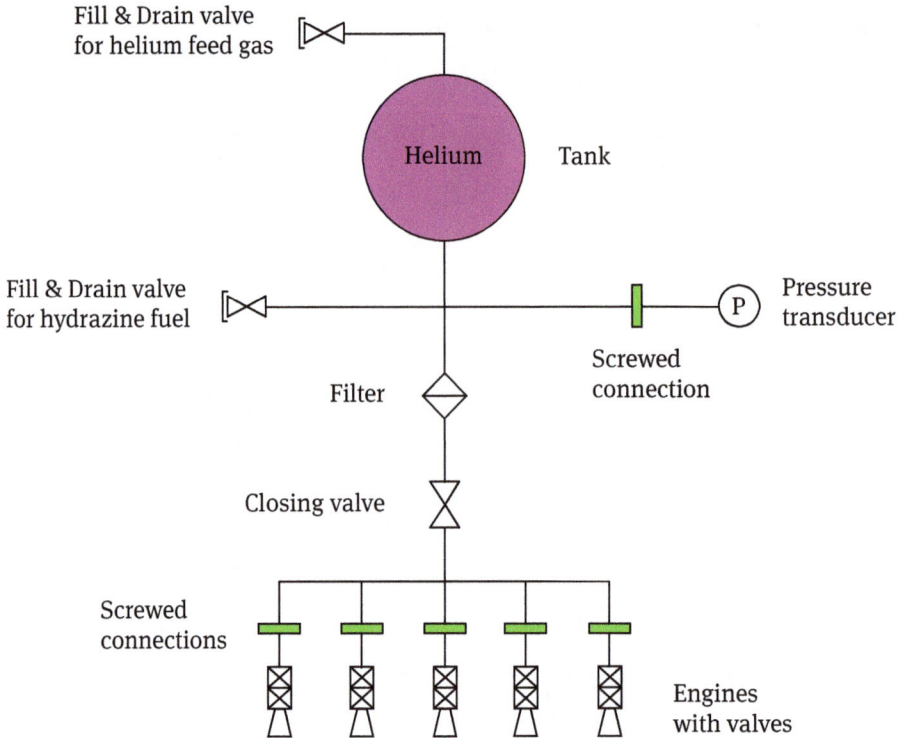

Figure 9.25: Basic structure of a monopropellant propulsion system.

9.4 Re-entry Bodies

A re-entry body is a satellite or a part of a satellite which is decelerated by the atmosphere of a celestial body at the end of a mission and being returned undamaged. This requires first a transitions of the re-entry body into a returning ballistic trajectory by an orbital manoeuvre. An additional braking system for damping hard landings is required, if landing takes place onshore (e. g. with Soyuz capsules). The American Mercury, Gemini, and Apollo capsules did not require this system, since splashdown was always carried out in the Pacific Ocean. The following main groups of re-entry bodies are distinguished and further re-entry systems are shown in Figure 9.26.

– Ballistic systems alone (in general, for military purposes)
– Disposable systems (landing capsules)
– Re-usable systems (Space Shuttle)
– Crew Rescue Vehicle (CRV)

Re-entry vehicle

Soyuz landing module
(Ballistice vehicle)

Dream Chaser
(Aerodynamic vehicle)

TSTO with re-usable upper stage

Expendable 1st / 2nd stage
(or parachute recovery)

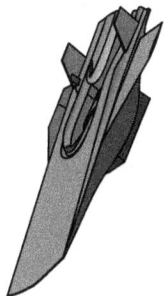

Space Operations Vehicle

Space Transportation System

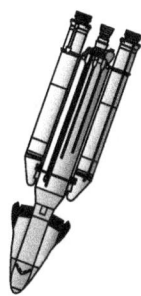

Hermes on top Ariane 5

Fully re-usable

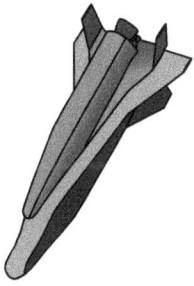

FSSC - 9

X37B

Sänger II

SSTO vehicle

DC - X
(Rocket – VTOVL)

Venture Star
(Rocket – VTOHL)

Skylon
(Air-breathing vehicle – HOTOL)

Figure 9.26: Typology of re-entry systems (not to scale).

9.5 Single-Stage-To-Orbit Vehicles

DC-X

The DC-X, short for Delta Clipper Experimental, was an unmanned single-stage-to-orbit (SSTO) vertical take-off and vertical landing (VTOVL) vehicle built by McDonnell Douglas Corporation on behalf of NASA and the Strategic Defense Initiative Organization (SDIO) of the U.S. State Department of Defense. In August 1993, the DC-X launch vehicle had its first lift-off at White Sands Missile Range in New Mexico. The experimental DC-X utilised liquid hydrogen (LH2) as fuel and liquid oxygen (LOX) as oxidiser. The prototype never reached orbital altitudes or velocity but instead a test feasibility of both suborbital and orbital re-usable launch vehicles using the VTOVL concept. The DC-X technology was taken over by NASA in 1996, and upgraded to improved performance to create the DC-XA, named Clipper Advanced, which was launched in May 1996 and in June 1996 for the last time.

Venture Star

The re-usable orbital spaceplane Venture Star was a single-stage-to-orbit (SSTO) vertical take-off and horizontal landing (VTOHL) vehicle proposed by Lockheed Martin Corporation. The spaceplane was intended to be the first commercial vehicle that would launch vertically but land horizontally. The spaceplane was equipped with two J-2S Linear Aerospike engines. In contrast to the traditional bell shaped rocket engines, Linear Aerospike engines maintaining their aerodynamic efficiency over a wide range of altitudes. The spaceplane was fueled with LH2/LOX as fuel and oxidiser. The project was cancelled in 2001 because of failures on the composite material of the fuel tank during pressure testing of the sub-scale X-33 technology demonstrator of the Venture Star.

Skylon

The unpiloted Skylon is a re-usable single-stage-to-orbit (SSTO) air-breathing spaceplane. Currently, the spaceplane is under development at Reaction Engines Limited in the United Kingdom. Designed for horizontal take-off and landing (HOTOL) the liquid hydrogen-fuelled spaceplane will use its air-breathing modus up to 28 km altitude using atmosphere's oxygen before switching to the pure rocket mode using the onboard liquid oxygen supply to reach orbit. The vehicle is capable to place a payload of approximately 12 t into low Earth orbit. The fuselage of the spaceplane contains a payload bay and 250 t of usable ascent propellants, 220 t of LH2 and 30 t of LOX. Additional to the main tanks there are cryogenic tanks which feed the orbital manoeuvring system, the reaction control thrusters, and the fuel cell power supply.

The spaceplane is powered by two Synergetic Air-breathing Rocket Engines (SABREs) that can operate in both air-breathing and rocket modes. Each SABRE consists of two Skylon Orbital Manoeuvring Assembly (SOMA) engines producing a four nozzle cluster for each SABRE. The testing of the key technologies was successfully completed in November 2012. First ground-based engine tests are planned for 2019 and an unmanned proposed test flight is scheduled for 2025.

9.6 Fully re-usable Vehicles

FSSC-9

The FSSC-9 concept vehicle was one of eight system studies which were eventually chosen for detailed design studies in the course of the Future European Space Transportation Investigations Programme (FESTIP). The programme was started in 1994 by the European Space Agency (ESA) to establish an approach to re-usable launchers. A wide range of different single-stage-to-orbit (SSTO) and two-stage-to-orbit (TSTO) concepts with different launch and landing modes and pure rocket and air-breathing engines were considered. The FSSC-9 vehicle has a VTOHL/TSTO configuration with parallel staging and cryogenic rocket engines. Booster and orbiter are fully re-usable and return to the launch site after mission completion.

X37B

The unmanned Boeing X37B is a single-stage-to-orbit (SSTO) vertical take-off and horizontal landing (VTOHL) military spaceplane, also known as orbital test vehicle (OTV). The U.S. Air Force officially stated that the X37B is a re-usable robotic orbital test vehicle for a number of advanced structural, propulsion, and operational technologies. Four OTV missions have been carried out for the Air Force. All of the missions were a secret. On 20 May 2015, the X37B OVT-4 spaceplane was launchend on top of an Atlas 5 rocket from Cape Canaveral Air Force Station. As of December 2015, the fourth X37B mission was still under way. A Hall thruster experiment is supposed to be conducted to improve thrust for manoeuvring of an advanced extreme high frequency (AEHF) military communications satellite.

Sänger II

The Sänger II spaceplane is a cancelled two-stage-to-orbit (TSTO) vehicle with a hypersonic air-breathing 1st stage with liquid hydrogen (LH2) as fuel and a delta winged 2nd rocket stage (Horus) propelled with LH2/LOX. Between 1985 to 1994, the design of the spaceplane was developed by Messerschmitt-Bölkow-Blohm (MBB), Germany, using the horizontal take-off and landing (HOTOL) concept. The plans originated from a Sänger I design at Junkers factories in West-Germany between 1961 to 1974. The 1st stage of the spaceplane would cruise on turbo-ramjets with a velocity of Mach 4.4 releasing the 2nd upper stage (manned/payloaded Horus upper stage or unmanned Cargus upper stage) to orbit after accelerating to Mach 6.

9.7 Re-entry Vehicles

Soyuz landing capsule

The Soyuz landing capsule is a ballistic re-entry body and part of the Soyuz spacecraft launched on top of a Soyuz rocket. The first flight of a Soyuz rocket was on 28 November 1966. The first manned flight into space was on 23 April 1967. The spacecraft was first used to carry cosmonauts to and from the Soviet space stations, Salyut and Mir. Today it is used to carry three crew members and supplies to and from the ISS. One spacecraft is always docked to the ISS in case of emergency.

The spacecraft is built of three modules. An orbital module is the upper part of the spacecraft. It carries the equipment necessary to dock with the ISS. A service module is the lower part of the spacecraft. It transports the telecommunications and attitude control equipment and the coupling of the solar panels. The descent or re-entry module is in the middle of the spacecraft. This section re-enters Earth's atmosphere and uses parachutes and small rocket engines to slow down.

Dream Chaser

The Dream Chaser spacecraft is a re-usable aerodynamic vehicle for manned suborbital and orbital flights developed by the American Sierra Nevada Corporation (SNC). The company was selected by NASA as third contractor for the Commercial Resupply Services (CRS) programme for delivering cargos, supplies, and astronauts to the ISS. Together with two other contractors, SpaceX and Orbital ATK, which already carry cargo to the ISS, SNC's Dream Chaser is capable to carry seven astronauts to and from the space station. A cargo variant of the spacecraft with an expendable cargo portion containing solar panels has been proposed for the next phase of NASA's programme for cargo re-supply of the ISS.

9.8 Expendable 1st / 2nd Stage Vehicles

Space Operations Vehicle

The unpiloted Space Operations Vehicle (SOV), referred to as the 'Quicksat', is an air-breathing re-usable launch vehicle concept. Being a responsive military spaceplane with aircraft-like operability it is capable to perform a broad range of orbital or sub-orbital military missions. The vehicle is also capable to deploy the Space Maneuver Vehicle (SMV) and the Common Aero Vehicle (CAV). While the CAV is performing hypersonic global strike missions by a strike mission configuration the SMV can only perform servicing operations like cargo delivery by a cargo delivery configuration to low Earth orbit (LEO) and polar orbit. A space-access configuration with a SMV is shown in Figure 9.26.

Hermes

Hermes was a project of the French Centre National d'Études Spatiales (CNES) in 1975, and later by the European Space Agency (ESA). The Hermes spacecraft would have been launched using an Ariane 5 rocket and was part of a manned space flight programme. The Hermes spacecraft would consisted of two parts. A cone-shaped resource module at the rear of the vehicle, which would jettisoned before re-entry, and the spaceplane itself which would re-enter Earth's atmosphere.

The spaceplane was designed to take three astronauts and pressurised payload of 3 000 kg into orbits of up to 800 km altitude. In 1992, the project was cancelled due to cost overrun and performance goals not achieved.

Space Transportation System

The greatest advance in re-entry technology was achieved with the American Space Shuttle. The preparation of the orbiter for re-entry starts 4 h before touchdown. The procedure begins with a de-orbit burn at an altitude of 282 km above ground on the opposite side of the Earth from Kennedy Space Center (KSC), flipping around and firing the main engines against the orbiter's heading. This slows the orbiter down for atmospheric re-entry over the Indian Ocean off the coast of Western Australia. Then the westernmost flight path of the orbiter proceeds across the Pacific Ocean to the Baja California Peninsula, across Mexico and Texas, out over the Gulf of Mexico, and on to the west coast of Florida to proceed across the centre of the state of Florida (Figure 9.27). The landing procedure of the orbiter is shown in Figure 9.28.

Figure 9.27: Easternmost and westernmost ground tracks of the orbiter's approach to KSC. Credit: NASA, enlarged display [4].

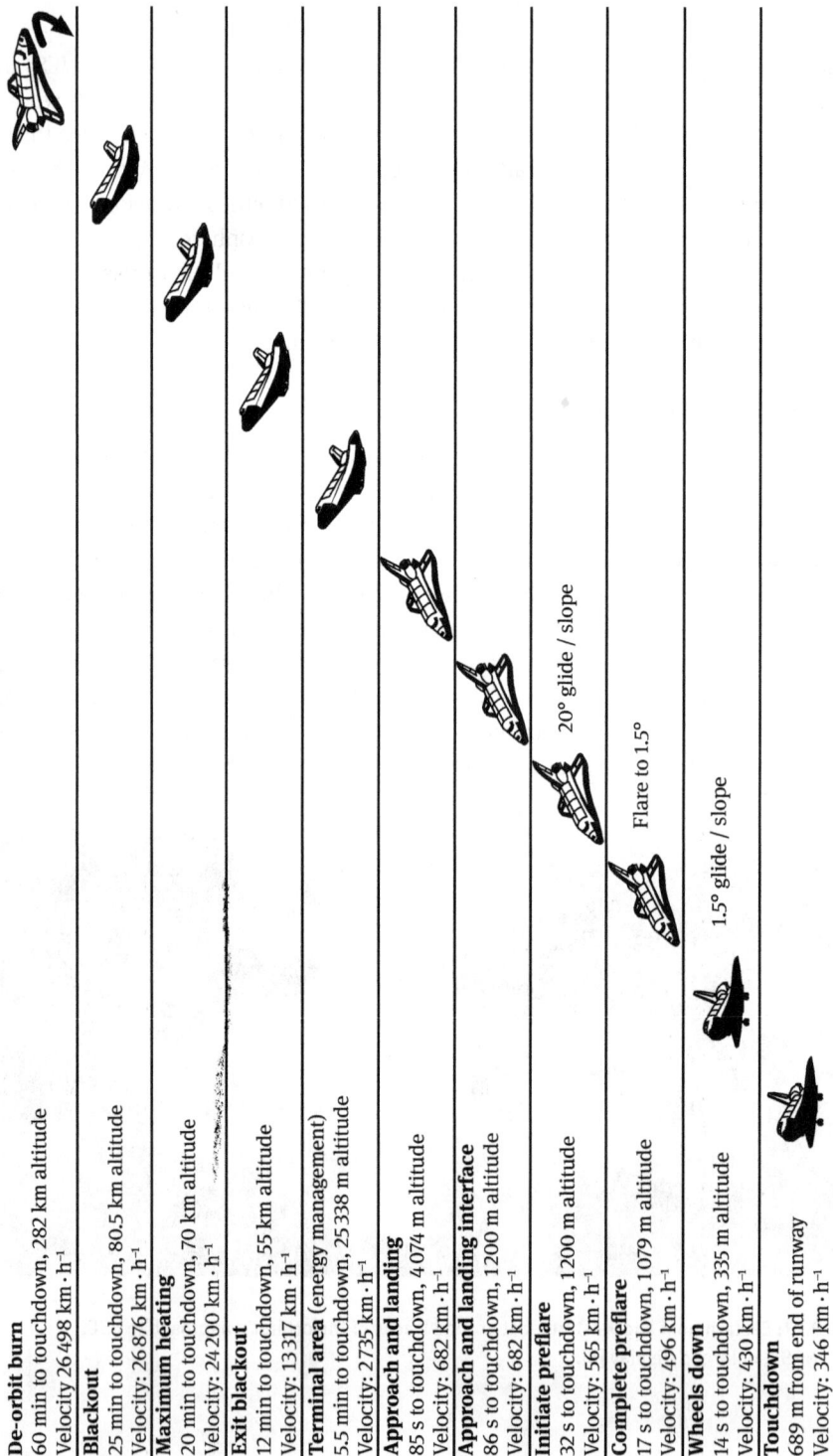

De-orbit burn
60 min to touchdown, 282 km altitude
Velocity 26 498 km · h⁻¹

Blackout
25 min to touchdown, 80.5 km altitude
Velocity: 26 876 km · h⁻¹

Maximum heating
20 min to touchdown, 70 km altitude
Velocity: 24 200 km · h⁻¹

Exit blackout
12 min to touchdown, 55 km altitude
Velocity: 13 317 km · h⁻¹

Terminal area (energy management)
5.5 min to touchdown, 25 338 m altitude
Velocity: 2 735 km · h⁻¹

Approach and landing
85 s to touchdown, 4 074 m altitude
Velocity: 682 km · h⁻¹

Approach and landing interface
86 s to touchdown, 1 200 m altitude
Velocity: 682 km · h⁻¹

20° glide / slope

Initiate preflare
32 s to touchdown, 1 200 m altitude
Velocity: 565 km · h⁻¹

Flare to 1.5°

Complete preflare
17 s to touchdown, 1 079 m altitude
Velocity: 496 km · h⁻¹

1.5° glide / slope

Wheels down
14 s to touchdown, 335 m altitude
Velocity: 430 km · h⁻¹

Touchdown
689 m from end of runway
Velocity: 346 km · h⁻¹

Figure 9.28: Landing procedure of the Space Shuttle orbiter

Figure 9.29: Space Shuttle *Discovery*, STS-131 lands at KSC on 20 April 2010. Credit: NASA [17].

The orbiter hits the atmosphere at an altitude of approximately 122 km above Earth. During the so-called entry interface the orbiter will begin to encounter the atmospheric effects. The orbiter's engines are off and it flies like a glider. The aft steering jets, as part of the reaction control system (RCS), controlled the orbiter's orientation. During descent the orbiter flies like an aircraft. The wing flaps and the rudder become active as the air pressure builds up. When the aerosurfaces become useable, the steering jets turn off automatically. In order to use up excess energy, and thus to slow down further, the orbiter performs a series of four rolling over manoeuvres, so-called steep bank manoeuvres, by as much as 80 degrees to one side or to the other. As a results of these steep banks the orbiter's track to landing appears as an elongated letter "S".

While moving through the atmosphere, the orbiter is faster than the speed of sound. The sonic boom can be heard over inhabited spaces along the flight path and will get louder as the orbiter looses altitude. When the orbiter eases off below the speed of sound the pilot takes manual control approximately 5 min before touchdown. As the orbiter nears KSC's Shuttle Landing Facility (SLF) and aligns with the runway, the orbiter performs a steep descent at an angle as much as 20° down the horizontal. Approximately 15 s before touchdown the pilot raises the nose of the orbiter to slow down the rate of descent to prepare for touchdown.

For the sake of completeness, it should be mentioned that a drag chute is deployed after touchdown to reduce the strain on the braking system and decreasing the landing distance (Figure 9.29). Step-by-step instructions from take-off to orbit to re-entry are given in the Space Shuttle Operator's Manual[16] in more detail. Moreover, the Shuttle Crew Operations Manual[17] is a comprehensive treatise on each Space Shuttle system and every phase of a generic Space Shuttle mission. The document contains condensed informations from a large number of Space Shuttle publications.

9.9 Questions for Further Studies

1. What space vehicles do you know?
2. How many propulsion stages were used for the Saturn V / Apollo system for Moon flights during the 1960s and 1970s? What for they were mainly used?
3. Are there any objects the Moon has captured in orbit?
4. What happend with ascent stage of the Apollo lunar module after ascending from the Moon and docking to the command and service module?
5. What had been done to solve Saturn V pogo oscillation problems?
6. Why the first and second stage of the Saturn V rocket were susceptible to pogo oscillations?
7. What is the purpose of the black-and-white pattern of some rockets?
8. What are the reasons to put the oxygen tank above or below the fuel tank for a given stage?
9. Why are there such differences in the number of engines on launchers?
10. What are turboramjet engines?
11. Explain the similarities and differences between a ramjet and a scramjet.
12. What are the differences between a normal bell-nozzle rocket engine and a linear Aerospike engine?
13. What is the purpose of the flame trench below the launch platform of the Space Shuttle?
14. What measures reduce the velocity of the Space Shuttle orbiter during landing approach?
15. What is the name of the first successor programme for the American Space Transportation System (STS)?

---

16 Joels KM, Kennedy GP, Larkin D. The Space Shuttle Operator's Manual. Ballantine Books, New York, 1982.
17 Sterling MR. Shuttle Crew Operations Manual. USA007587, Rev. A, CPN-1. United Space Alliance, LLC, Houston Texas, 2008.

10 Projects and Payloads

In general, space technology is carried out in project phases. Usually, there are six distinct project phases:

- Phase A: Conception
- Phase B: Definition
- Phase C: Development
- Phase D: Production
- Phase E: Operation
- Phase F: Final use

Projects roughly fall into three groups concerning their contractors.

- Commercial projects, funded by private parties or industry
- Scientific projects, funded by space agencies (with taxes)
    - NASA (United States)
    - ESA (Europe)
    - CNES (France)
    - NASDA (Japan)
    - ISRO (India)
- Military projects, funded by governments (with taxes)
    - Ministries of Defence
    - Procurement Offices for Army, Air Force, and Navy

For the sake of completeness, it must be mentioned here that in the university sector and in the amateur radio sector also small satellite projects are carried out to a considerable extent. These non-revenue 'companions' use the opportunity to fly with as second payload during larger missions once a supplier offers free lift capacities. The International Space Station (ISS) is currently the largest and most expensive project of space applications with a planned investment budget of approximately 100 billion euros.

The European space industry and others contribute to the ISS with the Columbus space laboratory and the Automated Transfer Vehicle (ATV). Until 2011 the Space Shuttle and the Russian Soyuz capsule transport crew members to the ISS and back to Earth. The ISS remains in operation until 2024. At the beginning of 2014, NASA confirmed that the U. S. Government communicated the extension of funding by 2024. In technical terms, the ISS is actually fully operational by the year 2028.

Figure 10.1: ISS photographed from Space Shuttle Atlantis, STS-132. Credit: NASA , rotated [18].

Figure 10.2: ATV Soyuz TMA-15M undocking from the ISS on 11 June 2015. Credit: NASA [19].

10.1 Commercial Projects

Since the beginning of space flight several branches of engineering meanwhile developed to economic independence by commercialisation. The applications included are listed below.

- Telecommunications
  - Radio and television
  - Fixed-network or mobile telephony worldwide
  - Electronic mail
  - Customers are Intelsat, Eutelsat, Astra SES, GE Americom etc.
  - Projects
    Symphonie (Germany / France, 1974)
    TV-SAT (Germany, 1987) and many others
- Weather observation in the visible, infrared and radar wavelength range
  - Local meteorological observation
  - Thunderstorm surveillance
  - Worldwide climate prediction
    Customers are Meteosat and many others
  - Projects
    Meteosat
    Meteosat Second Generation (MSG)
    Meteosat Third Generation (MTG)
    Metop and many others
- Navigation
  - Localisation by land, sea, and air
  - Cartography
- Earth observation
  - Environmental monitoring
  - Marine observation (oceanic currents, temperatures, marine industry)
  - Agriculture and forestry
  - High-resolution image recording
- Launch vehicles
  - Delta, Atlas, Ariane, Zenit, Long March, GSLV, Proton, Soyuz, Vega etc.
  - Customers are Arianespace, Boeing, Sea-Launch and others
- Tourism

10.2 Scientific Projects

After the first military use, but before commercialisation of space flight, scientific projects have been established. The technical requirements for this are naturally high upon first investigation of the scientific environment. By no means exhaustive, some unmanned missions of past and present are listed in Table 10.1 below.

Table 10.1: Unmanned missions from past to present.

Project / Mission	Country	Year	Remarks
Sputnik 1	Soviet Union	1957	First artificial Earth satellite
Explorer 1	United States	1958	Earth science
Luna	Soviet Union	1959	Moon landing
Venera	Soviet Union	1961 – 1963	Venus orbit
Ranger	United States	1961 – 1965	Close-up images of the Moon's surface
Surveyor	United States	1966 – 1968	Soft landings on the Moon
Mariner	United States	1962 – 1975	Robotic interplanetary probes
Lunokhod	Soviet Union	1970	Robotic lunar rovers
Uhuru	Italy / United States	1970	First x-ray satellite
Viking	United States	1975 – 1976	Mars landing
Pioneer	United States	1972	Planetary exploration
Helios	Germany / United States	1975 – 1979	Solar system exploration
Voyager 1 and 2	United States	1977 to date	Studying the outer solar system
Exosat	Europe	1983	X-ray observatory
Vega 1	Soviet Union	1984	Flyby of Halley's Comet
Giotto	Europe	1985	Flyby of Halley's Comet
Hipparcos	Europe	1989	Mapping of stars
Magellan	United States	1989	Venus orbit
Galileo	United States / Europe	1989	Jupiter orbit
Rosat	Europe	1990	X-ray satellite
Ulysses	United States / Europe	1990	Polar Sun orbit
Hubble	United States / Europe	1990	Space telescope
ISO	Europe	1995	Infrared telescope
SOHO	United States / Europe	1995	Sun observation in L1
Mars Pathfinder	United States	1996 – 1997	Base station and roving probe on Mars
NEAR	United States	1996	First satellite around asteroid Eros
Cassini	United States / Europe,	1997	Saturn orbit and Titan lander Huygens
XMM	Europe	1999	X-ray satellite
Chandra	United States	1999	X-ray satellite
Artemis	Europe	2001	Technology carrier
Envisat	Europe	2002	Earth-observing satellite
Integral	Europe	2002	Gamma-ray astronomy
Mars Express	Europe	2003	Mars probe
Rosetta	Europe	2004	Comet probe
MESSENGER	United States	2004	Mercury probe
Venus Express	Europe	2005	Venus probe
New Horizons	United States	2006	Pluto-Kuiper Belt probe
Hershel/Planck	Europe	2009	Basic astronomy

Many of the scientific missions use special properties of environmental conditions in outer space. The absence of Earth's atmosphere, *inter alia*, offers the opportunity of optical observations outside the visible range, too. The Earth's atmosphere is opaque to different degrees for several wavelengths and hampered observational astronomy in the

- infrared range,
- ultraviolet range,
- x-ray range, and
- gamma ray waveband.

In basic research projects, it is becoming more and more common to use space applications (gravitational waves, cosmology, etc.).

Table 10.2: Manned missions.

Project / Mission	Country	Year	Remarks
Vostok	Soviet Union	1961	First man in space
Mercury	United States	1959 – 1963	First human space flight
Voskhod	Soviet Union	1964 – 1965	Human space flight
Gemini	United States	1961 – 1966	Human space flight
Apollo	United States	1969 – 1972	First man on the Moon
Skylab	United States	1973 – 1979	Scientific experiments
Space Shuttle	United States	1981	First re-usable space shuttle
Spacelab	United States	1983 – 1988	Scientific experiments
Soyuz	Soviet Union / Russia	1960 to date	Human spaceflight
Salyut	Soviet Union	1971 – 1986	Research / Military reconnaissance
MIR	Soviet Union / Russia	1986 – 2001	Microgravity research laboratory
ISS	International cooperation	1998 – 2024	Scientific research
Shenzhou	People's Republic of China	2003	First taikonaut in space
STS-135	United States	2011	Final mission of a space shuttle

10.3 Military Projects

Military requirements for more and more effective weapons were the driving force behind extensive funding of aerospace technology. In Nazi Germany, these requirements were taken into account by the construction of production plants at Nordhausen in the Harz mountains and at Peenemünde on the Usedom island. On the island of Usedom large launching facilities were operated, which were almost completely dismantled or destroyed after the Second World War. Today, remains of the pioneering days of space flight can be visited in a museum. After the end of the war, technicians and engineers were brought to the countries of the victorious allies in order to participate essentially in the development of space technology. Military aerospace have become an integral part of the strategies of world powers and emerging nuclear powers. The fields of application are very multifaceted.

Espionage and information

High resolution in optical observation of targets on Earth is achieved by extremely low-flying satellites. Observation can be hindered by camouflage and clouds. The optical observations are supplemented by radar and infrared image recordings. Interception of electronic information channels (for military purposes and increasingly used for industrial purposes) is growing in importance. Surveillance of nuclear explosions aboveground or in the atmosphere can only be reliably performed through reconnaissance by special orbiting satellites.

The working life of low-flying spy satellites, to a limit of 200 km altitude, lasts only a few days, since re-boosting has often to be performed due to increasing density of the residual atmosphere. After exhaustion of the propellant the satellite burns up in Earth's atmosphere within a few days. Some former Russian satellites of the Cosmos series used energy supply systems based on radioactive decay of isotopes. During burn up of these satellites, radioactive contaminations to varying degrees repeatedly occurred in a random, incalculable crash zone.

Navigation

Specific navigation systems are used for fire control of different types of weapons. The American Global Positioning System (GPS), officially designated as is a system which is available for civilian use to some extent, with restricted accuracy. But when needed, it is exclusively used by military. Originally, the system used 24 satellites and became fully operational in 1995.

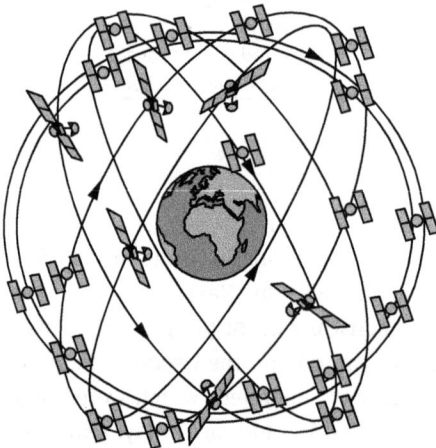

Figure 10.3: Global positioning constellation.

The development of software to determine satellite obits more accurately by NASA's Jet Propulsion Laboratory led to the development of the GPS Software System. The Global Positioning System is intended for pinpointing the three dimensional position of an object (latitude, longitude, and altitude) to approximately a metre of accuracy. The GPS is divided into the following three main parts.

- A constellation of at least 24 satellites placed in six orbital planes in a Medium Earth Orbit (MEO) with an inclination of 55° from the equator at an altitude of approximately 20 200 km. These satellites orbiting the Earth every 12 h.

- Satellite control stations on Earth controlling, monitoring, and maintaining the GPS satellites.

- Receivers for users that process the navigation signals from the GPS satellites to calculate position and time.

Launcher systems for weapons

The first use of aerospace engineering on a large technical scale was to carry bombs to targets (London, Coventry, Antwerp) which were not reachable in another way. Today, any place on Earth is reachable by Intercontinental Ballistic Missiles (ICBM) in less than 45 minutes. Relocation of armoured launcher systems towards the potential enemy territory shorten the period of time by 20 minutes. The launcher systems are nuclear-armed or carry conventional explosives with single or multiple warheads. They are ready to launch from missile silos at ground, submerged submarines, and long-haul aircrafts in the air for short-term use.

The propulsion systems of such armoured launcher systems are almost exclusively operated by solid propellants, since short-term operational readiness 'at the touch of a button' using liquid propellants is only limitedly possible (limited storage life, long-lasting tanking procedures, etc.).

SDI and NMD

Today, whole strategies of defence and warfare are planned strictly confidential by means of aerospace technology. In the United States, these plans have become known as Strategic Defense Initiative (SDI) and National Missile Defense (NMD). There is a whole arsenal of possibilities to neutralise approaching missiles. Many of the tests cannot be performed entirely confidential, because of observations from outer space. However, details will not made available to the public, of course.

It would be extremely dangerous for world peace if political unstable, anti-democratic powers are able to gain possession of long-range guided missiles (e. g., North Korea with launching facilities in No-Dong). Up to now, the superpowers of the world have accumulated a weapon potential that the extinction of the whole species of the Homo sapiens sapiens will not appear unlikely in any future. Space technology shares ethical responsibility for this, too.

10.4 Questions for Further Studies

1. Explain the six phases of completion of a space project.
2. What does space projects can be grouped to? Give examples!
3. What does OSCAR and AMSAT stand for?
4. Give the names of the Russian and European equivalent to the American Global Positioning System (GPS).
5. What are the reasons to establish space-based satellite navigation systems?
6. What is the name and the task of the first orbiter of the American Space Transportation System program?
7. What is the purpose of the International Space Station?
8. Find out the names of the operational orbiters which were used to serve the International Space Station in the past.
9. How the International Space Station will be maintained in future after the decommissioning of the orbiters of the American Space Transportation System?
10. What is an Automated Transfer Vehicle?

11 Launch Sites

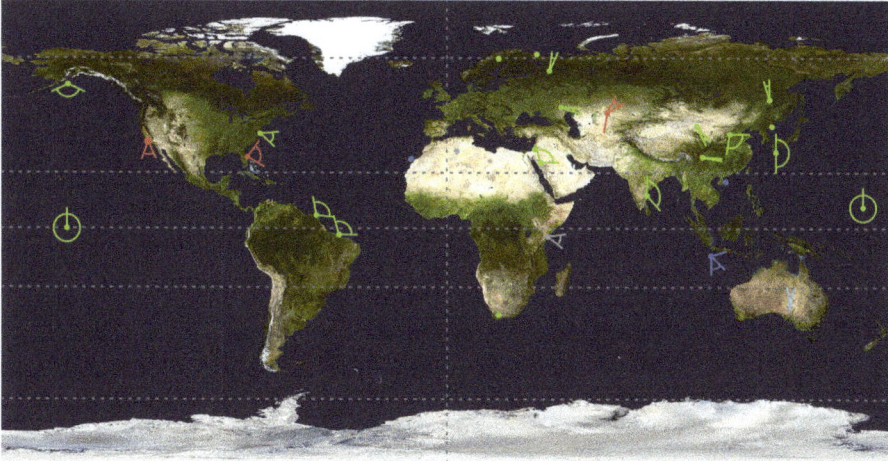

Figure 11.1: Current rocket launching pads with launch directions. Credit: NASA [4].

Since the beginning of large-scale use of rockets for space flight (Peenemünde on the island of Usedom, 1942), different launch sites were established and maintained. Details on the geographical longitude and latitude, permissible launch directions (Azimuth = 0 correlates to the North, Azimuth = 90 correlates to the East, etc.), rough data of world time zones, and initial entry of service of 30 launch sites are listed in Table 11.2. The distribution of launch sites is shown in the above Figure 11.1.

Table 11.1: Characteristics of current launch pads.

Characteristic	Location	Colour
Manned	Vandenberg, Kennedy Space Center, Baikonur	Red
Unmanned	Wallops Island, Kiruna, Barents Sea, Plesetsk, Juiquan, Svobodny, Kagoshima, Kodiak, Kourou, Palmachim, Kapustin-Yar, Taiyuan, No-Dong, Tanegashima, Sea Launch, Alcantara, Overberg Toetsbaan, Sriharikota, Xichang, Kwajalein	Green
Planned	Kiritimati, Gando AFB, Christmas Islands, Hainan	Blue
Historical	Hammaguir, San Marco, Woomera	Gray

In addition to earlier solely stationary facilities used, there are also mobile launch pads for space mission optimisation. Sea launch from a modified former oil platform and air launch from modified commercial aircrafts are available for commercial launches. Thus, the launch site is better adapted to the requirements with regard to geographical longitude and latitude.

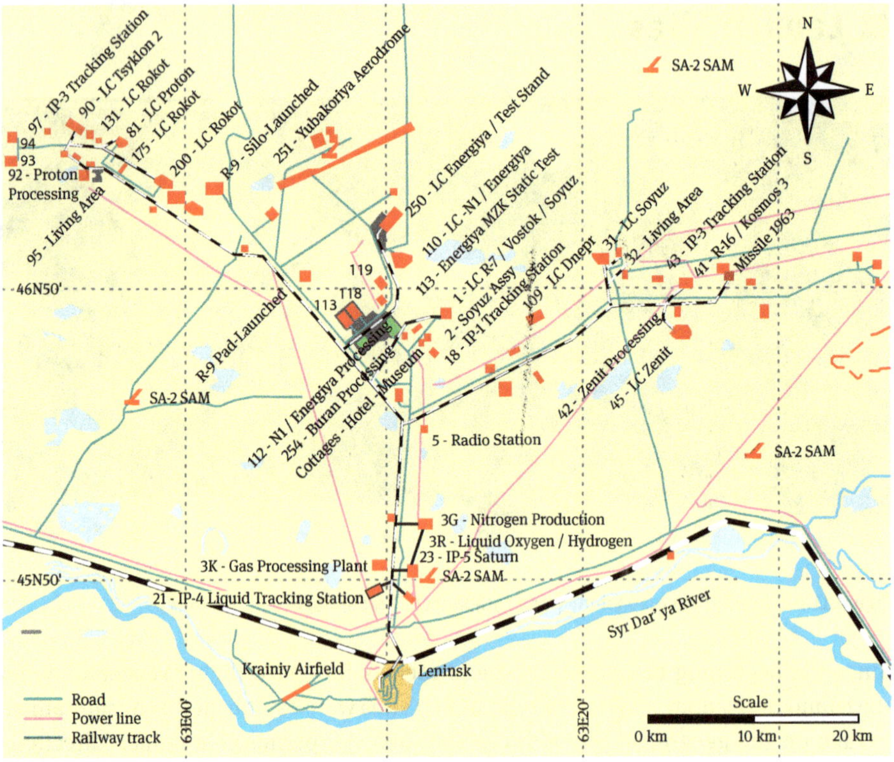

Figure 11.2: Site map of the Russian Baikonur spaceport. Credit: © Mark Wade, modified [20].

Sea Launch from a modified former oil platform and Air Launch from modified commercial aircrafts are available for commercial launches. Thus, the launching site is better adapted to the requirements with regard to geographical longitude and latitude. Launch from submarines (see Table 11.1) is rather ascribed to curiosities of aerospace than that none-military applications will thereby emerge in future.

11.1 Baikonur Spaceport

The largest spaceport is situated in the sparsely populated area of the Republic of Kazakhstan north of the settlement of Tyuratam. The Baikonur spaceport is operated there by the Russian Space Agency on an area of over 1000 km². The name Baikonur derives from a very distant city of the same name. Originally, the name was used as camouflage to confuse hostile observations by the Americans. The terrain is spaciously laid out to maintain large safety distances between several launch installations. After final assembly, the launchers are transported horizontally on rail tracks to different launch pads, refuelled and launched there.

Table 11.2: Launching sites.

Launching site	Country	Latitude / deg	Longitude / deg	Azimuth min	Azimuth max	Time zone	Service	Comments
Alcantara	Brazil	-2.20	-4.2	-17.0	90.0	-4	2000	
Anheung	South Korea	36.50	126.5			+8	2002	
Baikonur / Tyuratam	Kazakhstan	45.60	63.4	25.2	62.5	+5	1957	Also Azimuth = 193
Barents Sea	-	69.30	35.3				1998	Submarine
Christman Island	Australia	-10.29	105.37	135.0	180.0	+6	2000	Also Azimuth = 79
Gando AFB	Spain	28.00	-15.4			-1	2001	Gran Canaria
Hainan	China	18.50	111.0			+7	1999	
Hammaguir	Algeria (F)	30.90	3.1			0	1965	Decommissioned
Juiquan	Sinkinag, China	40.60	99.9	135.0	153.0	+7	1975	
Kagoshima	Japan	31.20	131.1			+8	1970	
Kapustin Yar	Kazakhstan	48.31	45.48	90.0	107.5	+4	1961	
Kennedy Space Center	Florida, USA	28.50	-80.5	35.0	120.0	-6	1958	
Kiritimati	Kiribati	1.80	-157.5			+12	2002	Plans of the NASDA
Kiruna	Sweden	68.00	20.0			0	1980	High-altitude research
Kodiak Space Center	Alaska, USA	57.50	-153.0	116.0	244.0	-10	1999	
Kourou	French Guiana	5.20	-52.8	-10.5	93.5	-4	1979	
Kwajalein	Micronesia (U.S.)	8.50	167.0	0.0	360.0	+11	1999	Reagan Test Site
No-Dong	North Korea	40.85	129.65			+8	1993	
Overberg Toetsbaan	South Africa	34.58	20.32			+1	1997	
Palmachim AFB	Israel	31.90	34.8			+1	1988	
Plesetsk	Russia	62.70	40.3	7.0	27.2	+3	1990	
San Marco	Kenya (I)	-3.00	40.0	82.0	130.0	+2	1972	Decommissioned
Sea Launch	-	0.00	-154.0	0.0	360.0	-11	1998	
Sriharikota	India	13.80	80.3	0.0	140.0	+4.5	1991	
Svobodny	Russia	51.50	128.5	-16.2	0.0	+8	1996	
Taiyuan	Gobi Desert, China	37.50	112.6	90.0	190.0	+7	1980	
Tanegashima	Japan	30.20	131.0	0.0	180.0	+8	1975	
Vandenberg	California, USA	34.70	-120.4	158.0	201.0	-9	1970	
Wallops Island	Virginia, USA	37.80	-75.3	85.0	129.0	-6	1965	
Woomera	Australia (GB)	-31.10	136.8	-10.0	15.0	+8.5	1965	Decommissioned
Xichang	China	28.25	102.0	94.0	104.0	+7	1985	

11.2 Kennedy Space Center

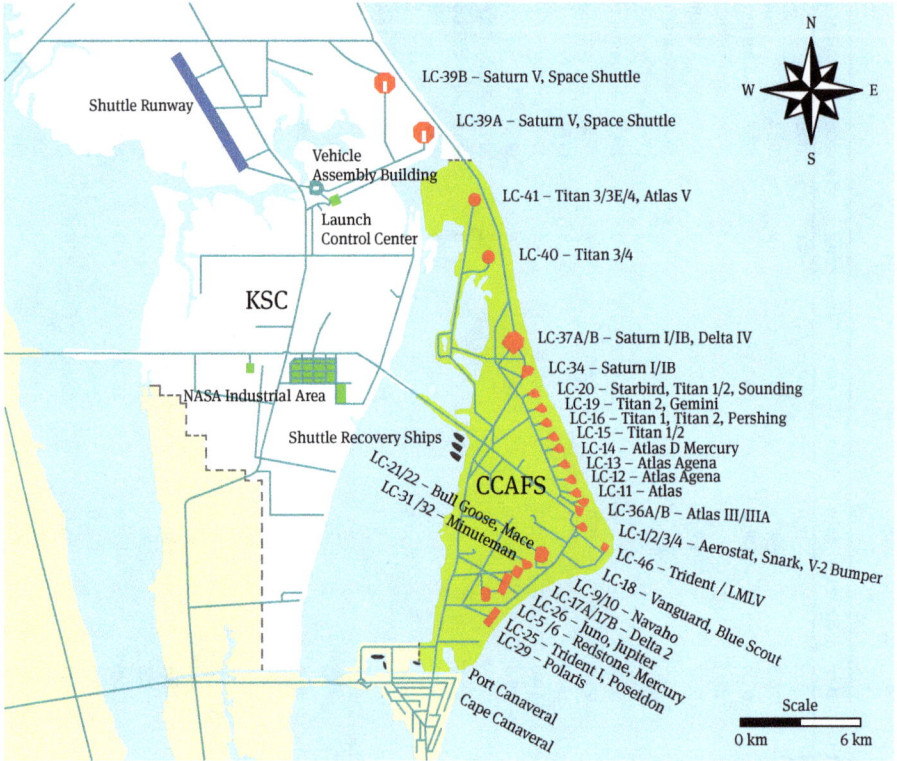

Figure 11.3: Site map of the Kennedy Space Center – Spaceport USA in Florida, USA. KSC is shown in white and CCAFS in olive.

The John F. Kennedy Space Center (KSC) is NASA's Launch Operations Center located on Merritt Island, Florida. The centre is north-northwest of Cape Canaveral on the Atlantic Ocean. The Launch Operations Center supports the Launch Complex 39 (LC-39A/B), which was originally built for the Saturn V rocket and the Apollo manned lunar landing program. After the Apollo programme has ended in 1972, the Launch Complex 93A/B has been used for Skylab in 1973, the Apollo-Soyuz Test Project in 1974, and for the Space Shuttle program during the years 1981 to 2011. There is also a Shuttle landing facility located to the north, which was used for most Space Shuttle landings.

Before the Kennedy Space Center was created for the Apollo manned lunar landing program in 1958, Cape Canaveral Air Force Station (CCAFS) became the test site for missiles in 1950. The first rocket launched at Cape Canaveral Air Force Station was a V-2 Bumper 8 rocket from Launch Complex 3 (LC-3). Of the launch complexes, three remain active until today with two planned for future use.

11.3 Guiana Space Centre

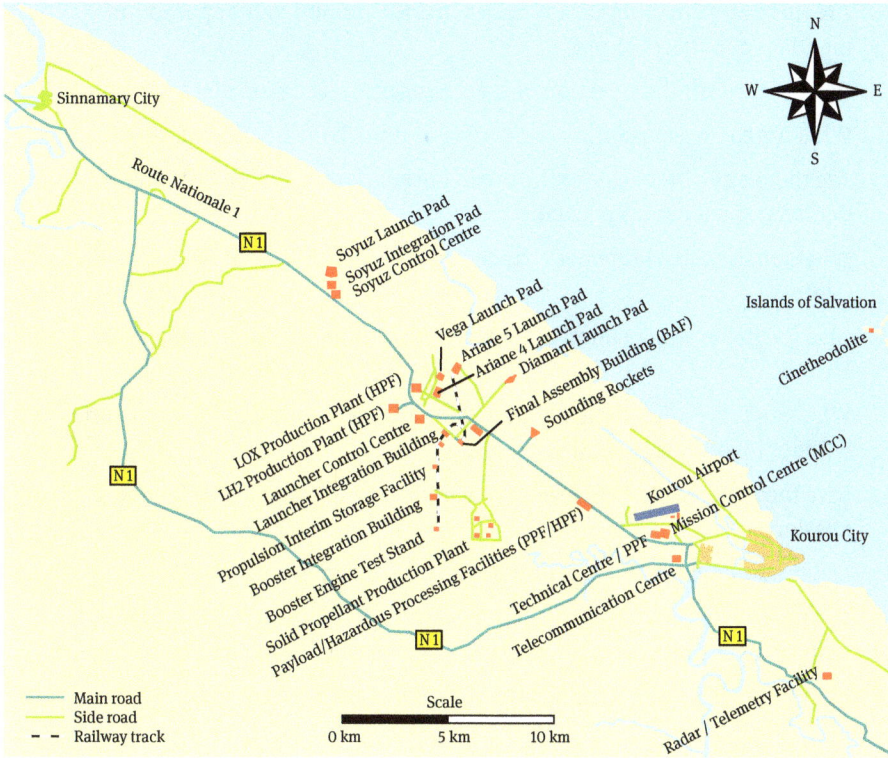

Figure 11.4: Site map of the Guiana Space Centre – European Spaceport near Kourou, Guiana.

The Guiana Space Centre (CSG from French abbreviation of Centre Spatial Guyanais) is located in French Guiana between the cities of Kourou and Sinnamary on the coast of the Atlantic Ocean in the northern part of South America. The Guiana Space Centre comprises five main complexes.

– CSG Arrival Area with sea and air ports managed by local authorities
– Payload preparation complex (EPCU) with
  – Payload Processing Facility (PPF)
  – Hazardous Processing Facilities (HPF)
  – Payload/Hazardous Processing Facilities (PPF/HPF)
– Upper Composite Integration Facility for each launch vehicle
  – Final Assembly Building (BAF) for Ariane 5
– Launch Sites for Ariane, Soyuz, and Vega
– Mission Control Centre (MCC or CDC – 'Centre de Contrôle')

11.4 Questions for Further Studies

1. Are there any other important spaceports in addition to the spaceports mentioned in this chapter?
2. Give reasons for the selection of the geographical location of a spaceport.
3. What are the reasons for creating a Soyuz launch facility in Kourou by Russia?
4. Are there any differences of the Soyuz rockets launched from Guiana Space Centre and Baikonur Spaceport?
5. What are the three launchers which conduct launches from Guiana Space Centre?
6. Give the name and location of a spaceport for launching sounding rockets in Europe.
7. What are the names of the German sounding rocket programmes with and without involvement of other countries?
8. Give the advantages and disadvantages of mobile launch pads for space missions.
9. List the advantages of launches operated by the international Sea Launch consortium.
10. Compare the advantages of sea-based over conventional land-based launch platforms.

12 Environmental and Boundary Conditions

In the space sector, a number of environmental conditions are important to consider when designing and using space products.

12.1 Environmental Conditions

Gas density

In general, the gas density in outer space is significantly smaller than under the best feasible vacuum conditions on Earth. Thereby, the gas density decreases with distance from Earth by several orders of magnitude. Close to the Earth, the gas pressure can reach the order of 10^{-6} Pa mainly depending on solar activity, and in the distance of the Moon the residual pressure is only 10^{-12} Pa. In some cases, the partial vapour pressure of different materials is therefore taken into consideration for long-term use. Different effects such as outgassing of plastics or losses due to sublimation in metals must not be ignored.

Temperature

The range of temperatures being found in outer space is considerably larger than on Earth, since the insulating effect of the Earth's atmosphere is lacking. Consequently, an equilibrium temperature of up to 500 K is adjusted during insulation. While passing Earth's shadow, the temperature could decrease to 150 K. The heat transfer onto the satellite occurs almost exclusively by heat radiation. Thermal conduction and convection are significantly smaller than under ground conditions on Earth, since particle density, as a carrier of heat, is very low. This can be expressed by expanding the mean free path of a particle up to the km-range (depending on temperature and particle density). For determination of radiant heat, the following radiation sources as black bodies should be taken into account in a first approximation.

- Sun with 5 600 K
- Earth with approximately 270 K
- Cosmic background radiation with 2.7 K

The radiation intensity is proportional to the fourth power of the temperature of a black body and decreases with the square of the distance from the radiation source.

A further particular environmental impact occurs upon re-entry of a vehicle (Space Shuttle, Apollo/Soyuz space capsule, or ICBMs) into the atmosphere of a celestial body. According to the Law of Conservation of Energy, the kinetic energy of a spacecraft, which is accelerated to high velocities, must be 'destroyed' again before landing or converted into another form of energy. Aerodynamic braking produces temperatures of 1 600 °C at the tip of the nose and at the leading edge of the vertical tail plane of the Space Shuttle. Much more higher temperatures accumulate on

- Intercontinental ballistic missiles (ICBM) by a steeper re-entry angle and a shorter braking time, and
- Apollo capsules by a higher approach velocity on the return from Moon.

These high temperatures result in partial dissociation and ionisation of the gas molecules within the vortex of the abruptly heated gas. Radio and radar waves cannot pass through this gas. A phenomenon called 'blackout' occurs regularly over a period of some minutes during re-entry of the spacecraft.

The uncontrolled crash of satellites without appropriate thermal protection always results in burning up of almost all components. Only massive and compact debris of the satellite (e. g., engines and structural components) hit Earth's surface and pose risks, despite of less frequency of occurrence which is considerably smaller than the daily impact of natural celestial bodies such as meteorites. Crashes of satellites carrying hazardous substances (e. g., radioactive substances within facilities for onboard energy supply) always result in an additional contamination of our environment.

Chemical composition of the environment

In the high atmosphere, the remaining existing oxygen is cleaved progressively by high-energetic radiation and high temperatures with increasing distance from the Earth. The atomic oxygen produced is a strong oxidising agent and attacks chemically all parts of a surface. Solar cells undergo an increased degradation associated with a decrease of efficiency during mission. Also emissions of corrosive gases and propellant leakages may have an unexpected influence on sensitive instruments of the payload by chemical reactions.

Radiation

Besides the thermal radiation already described, spacecrafts are almost unhamperedly exposed to existing x-rays, gamma radiation, and particle radiation. The level of penetration is different and depends on the properties of the radiation. The existing life on Earth is relatively well shielded from radiation by the atmosphere and individual molecules inside it (e. g., within the ozone layer). Similar effective protection against radiation in manned spacecrafts cannot be achieved by construction, since necessary massive shieldings would leave no space for payloads regarding today's launch capacities. Thus, American astronauts, Russian cosmonauts, and Chinese taikonauts will be exposed to high doses of radiation and known health risks during their mission in future. Also electronic components of unmanned satellites are sensitive to radiation. During peaking of sunspot activity, a complete failure of radio communications might occur.

Magnetic fields

The properties of magnetic fields are well known and must be also taken into account in designing spacecrafts and spacecraft's components. Close to the gas planets of Jupiter and Saturn, powerful magnetic fields exist. This means that planned gravitational manoeuvres (see Chapter 2.3.6) close to these planets are also avoided and a minimum distance have to be maintained depending on risk analysis.

Micrometeorites

A further risk for spacecrafts arises from the occurrence of periodically enhanced or irregular meteorite impacts. In the orbit, this risk is greater than on Earth (because of the Earth's protecting atmosphere). Every day, parts of satellites and their payloads were hit from different angles by minute meteorites with an energy in the range of 1 J. With increasing size of natural meteorites, artificial debris of satellites, upper stages of rockets and much more other things, the probability of being hit decreases. An element of risk which may lead to a total failure or loss of the whole mission always remains.

There is a moderately increased element of risk when moving through the planetary asteroid belt in the course of an interplanetary mission. Mostly, a thin foil in front of the satellite's part requiring protection 'destroys' (vapourises) the kinetic energy of the micrometeoroid on impact. Depending on the distance, there is a considerably higher risk during passage of a comet's tail. The Giotto comet probe lost its most important camera during flyby of Halley's Comet. If there is a breakdown of radio communications by an impact of a meteorite, the reason of the total loss cannot generally be identified afterwards.

12.2 Boundary Conditions

Weightlessness

Weightlessness, or rather microgravity, is a prerequisite for performing different processes in research, development, and manufacturing. Today, microgravity on Earth is established for a limited period of time in

– drop towers,
– parabolic flights, and
– sounding rockets.

For an unlimited period of 'free-fall', the International Space Station, which moves around Earth at approximately 400 km altitude in 'free-fall', is currently available.

At the drop tower a sample container is released to fall from the upper end of an evacuated tube. The period of 'free-fall' in the tube depends on the tower's height (falling zone). At the bottom of the tube, the sample container is stopped with multiples of acceleration due to gravity under controlled conditions by a braking device. Thus, a 125 m drop tube facilitates 5 s of weightlessness (microgravity). The period of 'free-fall' can be doubled, if the sample container is thrown upwards with higher technical efforts, then reaching the upper end and running downwards the tube a second time.

The period of 'free-fall' can only be extended by lengthen the drop distance using this procedure. One possibility is to perform more extended 'free-falls' onboard aircrafts, so-called parabolic flights. During these flights, the aircraft accelerates at low altitude and moves up to climb flight due to the laws of the oblique throw. Thus, the duration of microgravity is extended to approximately 40 s at a drop altitude of 2000 m. The existing external air drag must be compensated by the thrust of the engines. These counterforces limit the quality of microgravity, but it is sufficient for preparatory training of future astronauts.

The drop distance can be extended once again by using sounding rockets. These single-staged, mostly solid propelled projectiles were injected into an almost vertical ascent trajectory reaching ceilings of over 100 km. This results in periods of 'free-fall' in the minute range without greater air drag but with relatively low microgravity. The European Space Agency (ESA) also operates a large landing site for performing sub-orbital flights near Kiruna, Sweden, near the Arctic Circle.

Free-flight phases of man-made, low mass celestial bodies have unlimited periods of 'free-fall'. Such a facility is *inter alia* the International Space Station (ISS). The quality of microgravity is limited by the existence of the residual atmosphere and by irregular manoeuvres for re-boosting und orbit correction. Also movements of masses within the space station (astronauts, translational movable parts, and rotating parts) affect the quality of microgravity.

Fluids, e. g. propellants in the tanks, are spread irregularly in weightlessness. Therefore, particular devices are required to ensure propellant supply. The effect of convection also vanishes under microgravity. A burning candle is quenched by its own combustion products, if no fresh oxygen is supplied by air circulation or blowing. In humans, long-lasting microgravity results in demineralisation of the bones. This effect can be slowed down by additional special work-out, but cannot be eliminated entirely.

Space sickness

There is a phenomenon in manned space flight which is similar to sea sickness showing the same symptoms like dizziness, loss of appetite and even nausea. The precise reasons are not fully understood, but it is suspected that incorrectly tuned signals originating from the eyes and organs of equilibrium of the tympanum are not processed in the brain. This problem might become important for long-term stays in space stations or future space missions without artificial 'centrifugal gravity'. Many a spaceman returned early to Earth from a space station because of this difficulty. Even space missions have therefore been terminated early. Since individuals differ in their susceptibility to space sickness, a strict selection of space crew personnel is made before each mission to minimise this 'operating risk'.

Loads during launch

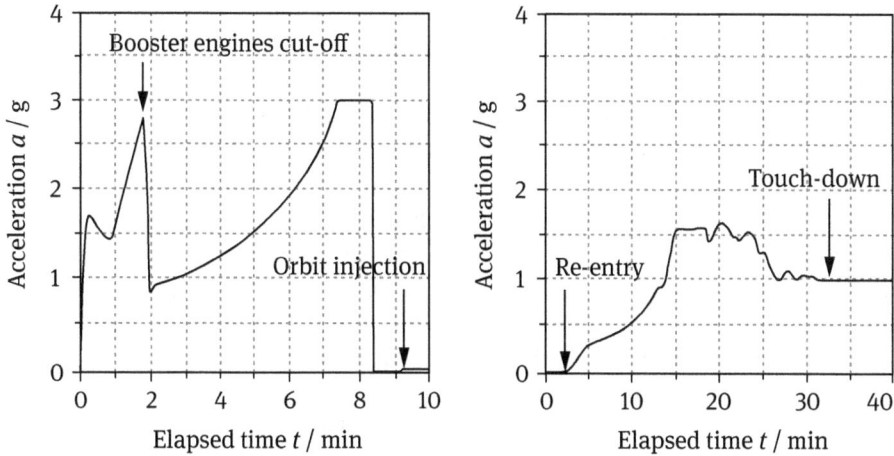

Figure 12.1: Load factor after lift-off and re-entry, respectively. Left: Load factor from lift-off to orbit injection. Right: Load factor from re-entry (approx. 122 km above ground) to landing. Credit: NASA.

On the way to weightlessness, considerable launching loads concerning the construction design of spacecrafts must be taken into account. The maximum of human resistance is an upper limit for manned spacecrafts, especially for the circulatory system and the eyes. Former launcher systems had a final acceleration shortly before the end of firing of up to nine times of the acceleration of gravity, shortly $9\,g$. In the former American Space Shuttle, this load was limited to a tolerable triple acceleration of gravity, $3\,g$, by reducing the thrust of the main engines (Figure 12.1). Returning from outer space and upon landing considerable static accelerations occur. In general, unmanned launcher systems are engineered for an increased acceleration up to $6\,g$ to reduce losses during ascent in the gravitational field of the Earth. These losses amount up to $2000\,\mathrm{m\cdot s^{-1}}$ (equivalent to a quarter of the final velocity in low Earth orbit). Directly after launch, the loss amounts to $9.8\,\mathrm{m\cdot s^{-1}}$ (equivalent to gravitational acceleration g) and reduces itself progressively with an increase of the centrifugal proportion which is directed against Earth's gravity. Due to firing of the thrusters during launch and by aerodynamic loads, severe vibrations were absorbed by structure and payload. Also acoustic loads (acoustic pressure) represent an enormous challenge to man and material. During stage separation or the release of protective covers and boosters, pyroshocks occur by explosive bolts and detonating cords with abrupt loads for assemblies and components (see Chapter 2).

12.3 Visibility of Satellites

Operation of artificial celestial bodies requires their availability and accessibility on certain locations on the quasi spherical surface of the Earth. These locations have to be determined firstly in a preliminary study. Radio communications and optical visibility without interruptions are only possible on a straight line. Therefore, it will be necessary to establish ground stations or relay satellites for sufficient long-term supervision and connection.

The geometric conditions of the observational horizons for a given observer in the orbit or on the surface of a celestial body are described in Chapter 2.3.2. Regarding the Earth, most satellites orbiting from west to east. The Earth also turns around its axis from west to east within 24 hours. Thus, geostationary satellites (GEO) are 'standing' above the equator, apparently stable at a specific position overhead for an observer on Earth. Satellites orbiting Earth below GEO move faster than the Earth, and therefore travel from west to east in the sky. While satellites above GEO move slower than the Earth, and therefore travel their orbits from sunrise in the west to sunset in the east.

Large satellites (e. g., ISS) are visible to the naked eye as the brightest objects (after the Sun and the Moon) in the sky after sunset in the evening or before sunrise in the morning, since the satellite is still being illuminated or is just already being illuminated by the sunlight, similar to a high flying aircraft. If the satellite then enters the shadow of the Earth or exits the shadow of the Earth, the satellite becomes abruptly invisible or visible (Figure 12.2).

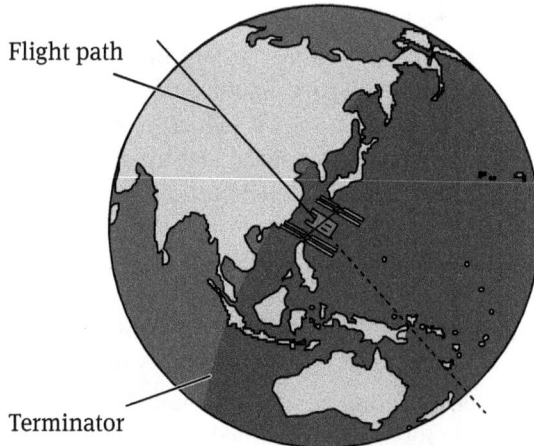

Figure 12.2: Visibility of the International Space Station over Taiwan after sunset (until the end of the solid line).

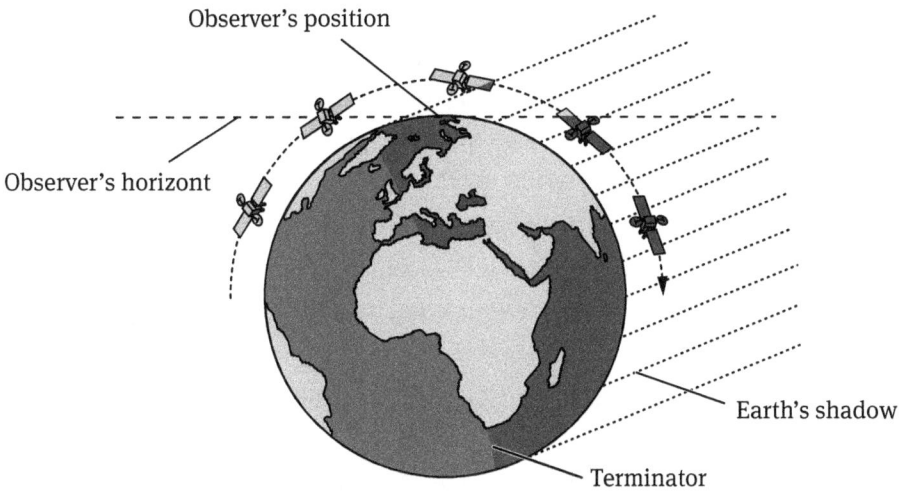

Figure 12.3: Visibility of a large satellite when flying over an observer at ground level.

Precise visibility data of the International Space Station (ISS) are continuously available from the heavens-above website[18] on the internet. Visibility data of the so-called Iridium flares are also available on this website. Since the Iridium satellite constellation has been positioned, bright light flashes can be observed in certain locations at certain times owing to reflection and bundling of the sunlight by the antennas of the satellites. This is a new form of light pollution which additionally interfere with optical (photographic) investigations in astronomy.

Launch windows

Appropriate times have to be calculated for each orbital manoeuvre (launching time of each mission or to initiate the return to Earth). These times regularly repeat for all periodically moving celestial bodies (synodic period). And thus, the Earth goes round once a day below the orbit of the Space Shuttle, the launch window has passed through only a few times per day for some seconds depending on orbit. For scheduled landing in Florida or California the firing of the OMS engines to initiate return must be accomplished within this window. Outside of this daily window further alternate landing sites are available for the Space Shuttle all around the world (*inter alia* also in Cologne, Germany).

Other launch windows (e. g., for interplanetary missions) only repeat after many years. Enormous work pressure on the ground personnel results from this. As an indirect cause, this pressure to succeed claimed more (unknown) fatalities than all space flights put together until now.

---

18 www.heavens-above.com

12.4 Questions for Further Studies

1. What environmental influences must be taken into account for reliable operation of a satellite in the geostationary orbit?
2. Which satellites can be observed from Earth with the naked eye?
3. By what means weightlessness can be realised?
4. What does the load factor represents in aeronautics?
5. When the load factor of an aircraft becomes $-1\,g$?
6. Why the load factor has the dimension of an acceleration?
7. What advantage a LEO satellite has when it is placed into certain polar orbits near the terminator?
8. What are the effects of spaceflight on the human body?
9. Space sickness is also called space adaptation syndrome (SAS). What is meant by 'syndrome'?
10. What kinds of orbital periods of objects around celestial objects are there?

13 Conclusions and Outlook

After the euphoria of moon flights in the sixties, the engineering works provided have never been repeated again to the same extend. In spite of high risks and costs space flight has established a niche existence in our today's technical civilisation and cannot be imagined from our everyday lives. In the long term, space flight will endeavour to concentrate on aspects of worldwide telecommunication including broadband internet worldwide and Earth observation. Also military necessities will provide a stable foundation for further developments of unmanned space flight.

In contrast, manned space flight is economically not profitable in the private sector to the present day. The American Space Agency (NASA) requires 3.2 billion euros per year of taxpayers' money for operation and maintenance of the space shuttle fleet (based on prices as of 2002). This means that the specific payload costs of manned systems are four to five-fold higher than of comparable 'disposable' launchers (Ariane 5, Atlas 5, Delta 4). Only the Russian Federation as successor to the Soviet Union and China yet operate manned launch systems mostly for reasons of prestige. European plans (small shuttle Hermes on Ariane 5 launch vehicle) were rejected after the Challenger disaster. The first manned mission of the People's Republic of China was performed in 2003. It has been speculated that the Chinese will leave some footprints on the Moon for thousands and millions of years second to the Americans.

Few successors of the today's manned launch vehicles are currently under construction for cost efficiency reasons, both in Russia and in the United States. In the medium term, the non-economic operation of the International Space Station (ISS) will show that manned space flight has no economic future. The expected benefit from manned space flight is incommensurate with the costs which might be caused thereby. In a long term, private investors and taxpayers were prevented from supporting further development of this branch of astronautics and colonisation of outer space.

Scientific curiosity will continue to be a driving force for the development of space flight concepts of unmanned robotic spacecrafts. This thirst for knowledge has the following goals.

- Clarify open questions in astronomy and cosmology.
  (gravitational waves, black holes, quasars, structure of galaxies, etc.).
- Search for Dark Matter and Dark Energy.
- Search for the Theory of Everything (GU Theory of the four interactions).
- Search for the origin of the universe, the solar system, etc.
- Search for the origin of life.
- Search for Extraterrestrial Intelligence (SETI).

Large space flight projects with European participation were achieved in recent years or will be achieved in the following years.

- Returning of rock samples to Earth by Mars Sample Return Orbiter (MSRO).
- James Webb Space Telescope, space telescope of the second generation (not before 2018).
- Investigation of gravitational waves by LISA (not before 2020).
- High accuracy measurement of stars by Gaia (launched in 2013).
- Exploration of planet Mercury by BepiColombo mission (January 2017).
- Exploration mission to Jupiter and Ganymede (JUICE, not before 2022).
- Investigation for Dark Matter and Dark Energy (EUCLID, not before 2022).

In the commercial sector, the installation of the GPS-independent European navigation system Galileo is in the implementation phase. The first units are in orbit and finalisation is expected in 2018. Meanwhile, another Indian regional navigation system (IRNSS) is in operation in the short-term. In the military sector, spy satellites are launched to obtain observation results and confirmed information for several international conflicts. In future, space technology will be a driving force for science, innovation, and new products like the legendary Teflon pan (by mistake), solar cells and fuel cells in the past. In any discussion on the benefits of space flight these arguments should not be overlooked.

13.1 Commercialisation of Aerospace Industry

Since the turn of the millennium, international aerospace has been characterised by two significant changes. The progressive commercialisation of aerospace and the advance of the People's Republic of China (further referred to as China) into the group of the major space-faring nations.

In the United States numerous private investors were attracted by space flight after the retreat of the American Space Shuttle. Since 2012, the International Space Station (ISS) has been supplied by a Falcon 9 launch vehicle along with the Dragon cargo spacecraft of Space Exploration Technologies Corporation (SpaceX) and the Antares launch vehicle along with Cygnus cargo spacecraft of Orbital Science Corporation (OSC). In the medium term, a new Space Launch System (SLS) will provide launch vehicles for missions to the Moon and deep space missions.

In China all launcher activities will be moved to Wenchang on Hainan Island to avoid launches above populated areas in the future. In 2011 and 2012, China has overtaken the United States in the number of rocket launches with the Long March rocket family. It is planned to establish a space infrastructure for commercial and military launches as well as for manned missions up to the Moon. For this purpose, new launchers were developed under the designation Long March 5, 6, and 7.

Russia is moving launcher activities from Baikonur in Kazakhstan to Wostotschny in the south-east of Siberia. As successors of the Soyuz rocket, launchers of the Angara family are being developed.

13.2 Scenario for Manned Space Flight to Mars

After the first manned space flights to the Moon (six landings with 12 Americans between 1969 and 1972) a similar important highlight in space flight was achieved never again. The realisation of manned flight from Earth to Mars would be a further increase in space flight. The theoretical basis for this was described by *Hermann Oberth*[19] and *Walter Hohmann*[20] for the first time. For further information on this subject a short elaboration with admittedly humble opinions is described.

13.3 Fundamentals of a Manned Mission to Mars

Planet Mars is the outermost planet of the four inner planets of the Solar system. Mars has only half of the size of our planet Earth in diameter. The mass of planet Mars is only 10 % of Earth's mass. For an orbit around the Sun (one Martian year) 687 days = 1.9 years are required. A day on Mars takes 24 hours and 40 minutes. Therefore, it is only insignificantly longer than a day on Earth. On the consistently solid surface of planet Mars astronauts meet a gravitational accelaration of $3.7\,\mathrm{m} \cdot \mathrm{s}^{-2}$. This is approximately equivalent to one third of weight on Earth but the twofold value on Earth's Moon. Furthermore, Mars has a thin atmosphere (approximately 6 mbar) of carbon dioxide (approximately 95 %), nitrogen (2.7 %), argon (1.6 %), and oxygen (0.3 %). The temperatures vary between − 60 °C at night and 0 °C during the day. Two moons, Phobos at an altitude of approximately 6 000 km, and Deimos at an altitude of 20 000 km, orbiting the planet.

In principle, Mars can only be reached from Earth by means of the latest technology only near of its oppositions to Earth. These planetary configurations are repeated by the mean length of a synodic period every 780 days (= 2.1 years) or rather 760 to 810 days. In opposition, the Earth E runs between the Sun S and Mars M. As a result, a close approach to Earth take place between 54 to $101 \cdot 10^6$ km depending on time and location of opposition in space. These fundamental differences are based on the shape of the Martian orbit. Due to an eccentricity of 0.093 the Martian orbit must not be assumed as a circular path in contrast to Earth or Venus, where this error is small by this assumption. At perihelion, the closest

19 Oberth H. Wege zur Raumschiffahrt. Verlag von R. Oldenbourg, München and Berlin, 1929.
20 Hohmann W. Die Erreichbarkeit der Himmelskörper. Oldenbourg Verlag, München, 1925.

approach of the orbit, Mars approaches the Sun up to 206 million kilometres, whereas at aphelion, the farthest approach, the distance increases up to 248 million kilometres. Therefore, Mars has an orbiting velocity between $21\,980\,\mathrm{m \cdot s^{-1}}$ at aphelion, and $26\,490\,\mathrm{m \cdot s^{-1}}$ at perihelion. Our spacecraft, in its state of motion, must be kept going at this velocity to allow soft approach and landing.

Basically, four options are presented for space probe flights from Earth to Mars and return. These options differ significantly regarding to their duration of flight and propellant consumption. In principle, one thrust with a predetermined intensity must occur for changing trajectory, at least. The strength of acceleration is expressed by the velocity requirement v. This velocity is transferred directly to a mass ratio consisting of mass m_1 before thrust divided by mass m_2 after thrust by the rocket equation $v = c \cdot \ln\,(m_1 \cdot m_2^{-1})$ of *Konstantin Eduardovich Tsiolkovski*. From this, a propellant consumption of $m_1 - m_2$ can be calculated. In this equation the exhaust velocity c of the firing gases is assumed to be constant, and ln is the natural logarithm. The duration of the thrusts were assumed to be small compared to flight time. Upon completion of the orbit injection manoeuvre for departure from Earth (at E_1), the space probe travels, or rather descends, powerless in free-fall to its destination. On arrival at Mars (at M_2) a thrust must be executed once more to avoid disastrous collision. A lander is then required for landing and lift-off from Mars. Additional amounts of propellant are required for this. Return flight from Mars is basically the same as the approach to Mars with a manoeuvre for departure from the Martian orbit. Arrival at the Earth is carried out with different selected velocities depending on trajectory. Deceleration is not accomplished by engines but directly carried out by touching high-level layers of the Earth's atmosphere, so-called aerobraking, and landing on water or on terrain, such as Apollo or Soyuz.

13.4 Possible Trajectories to Mars

All trajectories start within Earth's gravitational field. There are two options for this: launch from Earth's surface or launch from Earth orbit (low Earth orbit or from the Moon). Since specifications for trajectories have to meet defined conditions (so-called launch windows), a launch from an already occupied orbit is very restricted (e. g. from International Space Station, ISS). Therefore, it is very unlikely. Ideally, the launch should be carried out in Kourou, French Guiana. Numerous flights must be carried out to an equatorial orbit at low altitude (approximately 250 km above Earth's surface) to provide the Martian spacecraft with necessary hardware, propellant and supplies. In the following flight options, a launch (at E_1) from this orbit is assumed at an altitude of 250 km, respectively. The position of the elliptic Mars orbit is represented roughly by means of the vernal equinox Y.

Option A

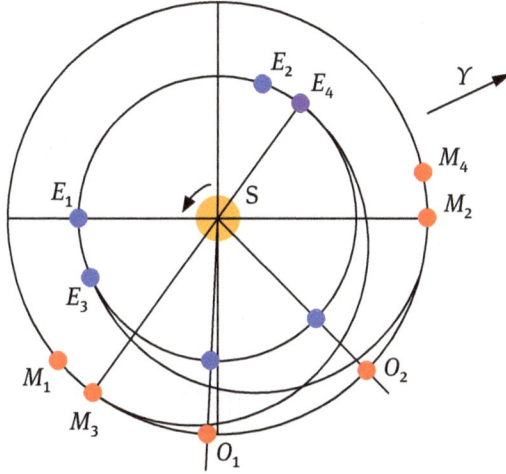

Figure 13.1: Two semi-elliptical trajectories for planet Mars.

Option A consists of two semi-ellipses (from E_1 to M_2, and from M_3 to E_4).

Start time E_1 :   3 to 4 months before opposition at O_1.

Arrival M_2 :   4 to 5 months after opposition at O_1.

Time remaining in the influence of Mars: 14 to 16 months.

Return flight M_3 :   4 to 5 months before opposition at O_2.

Arrival E_4 :   3 to 4 months after opposition at O_2.

Unbraked entry into the Earth's atmosphere with $11\,320\,\mathrm{m\cdot s^{-1}}$ to $11\,620\,\mathrm{m\cdot s^{-1}}$.

Total flight time: 2.5 to 3 years.

Velocity requirement: Four thrusts at E_1 ($3\,450\,\mathrm{m\cdot s^{-1}}$ to $3\,750\,\mathrm{m\cdot s^{-1}}$) according to distance of Mars), at M_2 ($1\,840\,\mathrm{m\cdot s^{-1}}$ to $2\,290\,\mathrm{m\cdot s^{-1}}$), at M_3 ($1\,840\,\mathrm{m\cdot s^{-1}}$ to $2\,290\,\mathrm{m\cdot s^{-1}}$) and at E_4 (aerobraking). Plus reserves for course corrections, this results in approximately $7\,500\,\mathrm{m\cdot s^{-1}}$.

Advantage: Less propellant consumption.

Disadvantage: Long flight time.

Option B

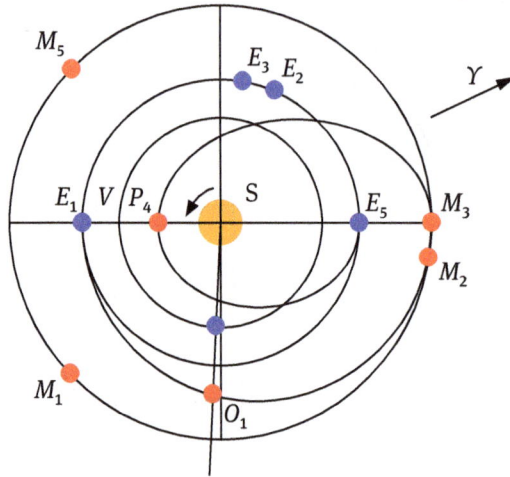

Figure 13.2: Three semi-elliptical trajectories for planet Mars.

Option B consists of three semi-ellipses (E_1 to M_2, M_3 to P_4, and P_4 to E_5).

Start time E_1 :	3 to 4 months before opposition at O_1.
Arrival M_2 :	4 to 5 months after opposition at O_1.
Time remaining in the influence of Mars:	Approx. 10 days.
Return flight M_3 to P_4 :	Approx. 6 months.

The lead of the Earth must be made up for close to the Sun.

Trajectory change manoeuvre P_4 :	Deceleration by approx. $3\,000\,\text{m} \cdot \text{s}^{-1}$.
Return flight P_4 to E_5 :	Approx. 4 months
Arrival E_5 :	Unbraked entry into Earth's atmosphere with $12\,600\,\text{m} \cdot \text{s}^{-1}$.
Total flight time:	Approx. 1.5 years.
Velocity requirement:	Five thrusts at E_1 ($3\,450\,\text{m} \cdot \text{s}^{-1}$ to $3\,750\,\text{m} \cdot \text{s}^{-1}$), at M_2 ($1\,840$ m·s^{-1} to $2\,290\,\text{m} \cdot \text{s}^{-1}$), at M_3 (approx. $5\,500\,\text{m} \cdot \text{s}^{-1}$), at P_4 (approx. $3\,000\,\text{m} \cdot \text{s}^{-1}$) and at E_5 (aerobraking). Plus reserves for course corrections, this results in approximately $14\,000\,\text{m} \cdot \text{s}^{-1}$.
Advantage:	Flight time shorter than for option A.
Disadvantages:	Higher propellant consumption than for option A. High thermal load close to the Sun at perihelion P_4.

Option C

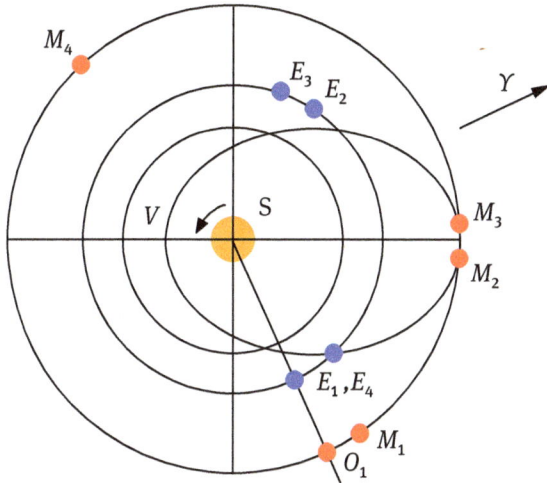

Figure 13.3: One complete ellipse for planet Mars.

Option C consists of one complete ellipse.

Start time E_1 :	Some weeks after opposition at O_1.
Arrival M_2 :	Approx. 4 months after opposition at O_1.
Time remaining in the influence of Mars:	Approx. 10 days.
Return flight M_3 to E_4 :	Approx. 8 months.
Arrival E_4 :	Unbraked entry into Earth's atmosphere with $16\,400\,\mathrm{m \cdot s^{-1}}$ to $23\,300\,\mathrm{m \cdot s^{-1}}$ (according to perihelion at P_4).
Total flight time:	Approx. 1 year.
Velocity requirement:	Four thrusts at E_1 ($8\,000\,\mathrm{m \cdot s^{-1}}$ to $15\,500\,\mathrm{m \cdot s^{-1}}$), at M_2 ($3\,700\,\mathrm{m \cdot s^{-1}}$ to $7\,200\,\mathrm{m \cdot s^{-1}}$), at M_3 ($3\,700\,\mathrm{m \cdot s^{-1}}$ to $7\,200\,\mathrm{m \cdot s^{-1}}$), and at E_4 (aerobraking). Plus reserves for course corrections, this results in approx. $15\,500\,\mathrm{m \cdot s^{-1}}$ close to Mars and $30\,000\,\mathrm{m \cdot s^{-1}}$ distant to Mars.
Advantage:	Flight time even shorter than for option B.
Disadvantages:	Even higher propellant consumption than for option B. Close to the Sun at perihelion P_4 with high thermal load. Very high re-entry velocity at E_4 with high load.

Option D

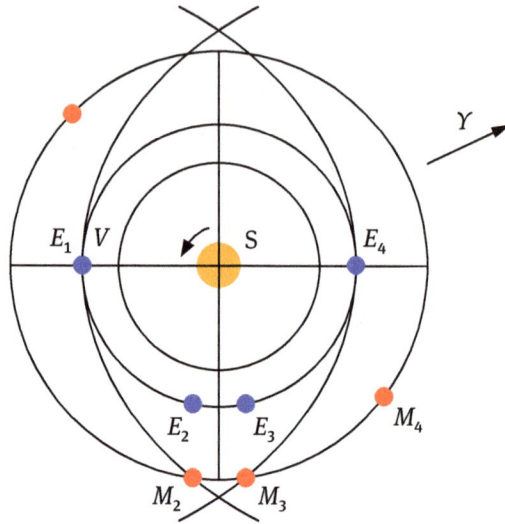

Figure 13.4: Two independant parts of ellipses or hyperbolas.

Option D consists of two independant parts of ellipses or hyperbolas, respectively.

Start time E_1 :	Approx. 3 months before opposition.
Arrival M_2 :	In the range of opposition.
Time remaining in the influence of Mars:	Approx. 10 days.
Return flight M_3 to E_4 :	Approx. 3 months.
Arrival E_4 :	Unbraked entry into Earth's atmosphere with $12\,000\,\mathrm{m \cdot s^{-1}}$ to $15\,000\,\mathrm{m \cdot s^{-1}}$.
Total flight time:	Approx. 6 months.
Velocity requirement:	Four thrusts at E_1 ($4\,560\,\mathrm{m \cdot s^{-1}}$ to $7\,220\,\mathrm{m \cdot s^{-1}}$), at M_2 ($8\,500\,\mathrm{m \cdot s^{-1}}$ to $15\,500\,\mathrm{m \cdot s^{-1}}$), at M_3 ($8\,500\,\mathrm{m \cdot s^{-1}}$ to $15\,500\,\mathrm{m \cdot s^{-1}}$), and at E_4 (aero-braking). Plus reserves for course corrections, this results in approx. $21\,600\,\mathrm{m \cdot s^{-1}}$ close to Mars and $38\,300\,\mathrm{m \cdot s^{-1}}$ distant to Mars.
Advantages:	Even shorter flight time than for option C. Lower re-entry velocity at E_4.
Disadvantage:	Even higher propellant consumption than for option C.

Flight times can be theoretically further shortened by using faster trajectories. These trajectories are characterised by a hyperbolic shape and a near-solar perihelion. The upper limit is thereby a straight line from Earth to Mars. In orbit mechanics, it is a hyperbola with an eccentricity asymptotically tending to infinity.

There is another propellant saving factor during approach to Mars. The lander is separated before braking manoeuvre at M_2 and heading directly to the surface of Mars by means of aerobraking without much of propellant consumption.

13.5 Landing on Mars

Landing on Mars is carried out from a near-ground orbit at an altitude of approximately 300 km above the surface of Mars. To initiate approach a braking pulse of $60 \, \text{m} \cdot \text{s}^{-1}$ is required. The deceleration of the lander is achieved by the atmosphere of Mars, the so-called aerobraking. For landing and following the line of approach, thrusts with a velocity requirement of approximately $50 \, \text{m} \cdot \text{s}^{-1}$ are accepted. The re-ascent of the lander requires $3590 \, \text{m} \cdot \text{s}^{-1}$ to reach orbit, $700 \, \text{m} \cdot \text{s}^{-1}$ are required for overcoming gravitational and atmospheric drag, and once again $60 \, \text{m} \cdot \text{s}^{-1}$ are required of docking to the orbiter. This results in a total velocity requirement of $4460 \, \text{m} \cdot \text{s}^{-1}$ for Mars which we want to assume for all options mentioned above. Therefore, this must be taken into account for our plannings as unchangeable.

The moon Phobos is fundamentally also conceivable for headquarters, but it is more a disadvantage for reasons of efficiency. Propellant consumption for landing can be minimised, if it is used as deep as possible within the gravitational field of Mars. The orbit should have an altitude of 300 km but the moon Phobos orbits at an altitude of 6000 km above the surface of Mars. The additional expenses amount to approximately $1690 \, \text{m} \cdot \text{s}^{-1}$ for headquarters on Phobos.

13.6 Plannings and Projects

Oppositions have to be determined first for a timeline of flight operations to Mars. The following table shows the next dates until 2031. Mars has its perihelion (closest distance to the Sun) at about 335° longitude. In 2010 an opposition at 336° longitude led therefore to a very close proximity to Mars that only happens once every 10000 years! There are two overlaps, X_1 and X_2, in options B and C with the Venusian orbit. The heliocentric longitude of Venus to Mars opposition is a crude estimate for a possibility of gravity assist manoeuvres, swing-by or fly-by, at Venus. These constellations are marked with an X and option B in parentheses (see Table 13.1). In option B there is a first possibility at approximately 190° after Mars opposition O_1 and a second one at about 190° after Mars opposition O_1. In option C the first possibility occurs at approximately 120° after Mars opposition O_1 and a second one at approximately 360° after Mars opposition O_1. Swing-by at X_1 can significantly reduce flight time while swing-by at X_2, shortly before arrival at the Earth, can reduce flight time only by a few days. Venus goes around the Sun in approximately 225 days and therefore it moves 1.6° of longitude daily. This allows Mars oppositions to be calculated.

Table 13.1: Mars oppositions for swing-by manoeuvres.

No.	Date of opposition	Longitude Mars	Longitude Venus	Swing-by possible at
1	27JUL18	304°	248°	
2	13OCT20	20°	104°	X_2 (Option B)
3	08DEC22	76°	283°	
4	16JAN25	116°	76°	
5	19FEB27	151°	222°	X_2 (Option B)
6	25MAR29	185°	6°	X_2 (Option B)
7	04MAY31	224°	161°	

Interestingly, there are no favourable constellations during the next oppositions for a swing-by manoeuvre at Venus according to option C (Table 13.1).

After the first soft Mars landing of the Viking probe in 1976, there have been numerous other successful, unmanned missions, but also unsuccessful missions. Manned space flight to Mars has been planned *inter alia* in the Aurora programme of the European Space Agency (ESA) as a vision between years 2025 to 2030. Unfortunately, Mars oppositions will then take place with increasing proximity to aphelion and as part of which, automatic rendezvous and docking tests will be investigated and performed in the Martian orbit (ExoMars mission, launched in March 2016). The Mars Sample Return Orbiter (MSR, launch not before 2020, see Chapter 13.6.2 for more details), an unmanned mission with a return of samples from Mars, will be performed for the first time. By means of progress of these prototypical missions a serious preparation for a manned space flight to Mars is suggested.

13.6.1 ExoMars Programme

The term ExoMars stands for exobiology on Mars. The programme's objective is to investigate if life ever existed on Mars. Together with the Russian space agency Roscosmos, the European Space Agency (ESA) will look for traces of life on planet Mars. The programme covers two missions to investigate the Martian environment and testing new technologies for future sample return missions. Lifting off from Baikonur spaceport in Kazakhstan on 14 March 2016, a Russian Proton M launcher carried the Trace Gas Orbiter (TGO) and the Schiaparelli entry, descent and landing demonstrator module (EDM) of the ExoMars 2016 mission into space.

Figure 13.5: Artist's impression of the Trace Gas Orbiter (TGO, left), the Schiaparelli module (EDM, centre), and the ExoMars rover (right). Credit: ©ESA – ATG medialab [21].

The ExoMars 2016 mission is carried out to detect and study atmospheric trace gases (such as methane, water vapour, nitrous oxides, and acetylene) by the TGO and to evaluate the performance of the lander and studying the environment at the landing site by the Schiaparelli module. The Trace Gas Orbiter (TGO) carries scientific instruments to detect and to characterise traces of gases in the Martian atmosphere. Carried by the TGO, Schiaparelli (EDM) will demonstrate the ability to perform a controlled landing on the surface of planet Mars. The EDM will separate from the TGO three days before the orbiter reaches the atmosphere of Mars and will coast to the planet in an hibernation mode to prevent battery depletion. After the atmospheric entry at an altitude of 122.5 km a parachute will be deployed at approximately 11 km altitude. The front heatshield separates and radar turns on at an altitude of 7 km. During the course of descent, monitoring data of all the key technologies of the mission will be transmitted back to Earth for subsequent flight reconstruction.

The parachute with the rear cover is jettisoned at 1.2 km altitude and the liquid propulsion system is activated to reduce velocity. At an altitude of 2 m, shortly before touchdown, the engines will be deactivated and the EDM lands in a freefall on a plain which is known as *Meridiani Planum*.

The mission's lifetime of the EDM is just a few days by using the excess energy of its primary batteries. A set of engineering and science sensors will continue to analyse the local environment after landing. Analysis of the environment is performed with the DREAMS package (Dust characterisation, Risk Assessment and Environment Analyser on the Martian Surface) consisting of a collection of sensors to measure the wind velocity and direction (MetWind), the humidity (DREAMS-H), the pressure (DREAMS-P), the atmospheric temperature close to the surface (MarsTem), the transparency of the atmosphere (Solar Irradiance Sensor, SIS), and the atmospheric electrification (Atmosperic Radiation and Electricity Sensor, MicroARES).

Table 13.2: Characteristics of Schiaparelli (EDM).

Characteristic	Value	Comment
Mass, complete	600 kg	–
Height	1.8 m	–
Diameter	2.4 m	With heatshield
	1.65 m	Without heatshield
Propellant	Hydrazine (N_2H_4)	–
Engines	9 x 400 N	3 clusters, pulse modulation
Power supply	Primary battery	
Communications	UHF	Linked with the ExoMars orbiter (2 antennas)

The ExoMars 2018 mission will deliver ESA's six-wheeled solar-powered rover and a Russian surface platform to the Martian surface. The rover provides mission capabilities like mobility on the Martian surface, ground drilling and automatic collection, processing, and distribution of samples to the onboard instruments. Launched into space by a Russian Proton rocket the rover and the surface platform will arrive planet Mars after a journey of 9 month.

Data from a number of nine instruments will help to perform a step-by-step exploration of the surface of Mars. This will be achieved by starting investigations at the metre scale and progressively going down to the sub-millimetre range, and finally, to the identification of organic compounds at the molecular level.

Panoramic Camera (PanCam)

The panoramic camera will perform digital terrain mapping of Mars. The camera system is a high resolution camera in combination with wide-angle cameras. They work together with the Infrared Spectrometer for ExoMars (ISEM) to locate surface targets for mineralogical investigations.

Infrared Spectrometer for ExoMars (ISEM)

The Infrared Spectrometer for ExoMars will assess the mineralogical composition of the surface to contribute to the selection of rock samples for further analysis by other instruments.

Close-Up Imager (CLUPI)

The close-up imager is a camera system to take high-resolution colour close-up images of rocks and ledges and drill fines and drill cores.

Water Ice and Subsurface Deposit Observation On Mars (WISDOM)

To provide informations on sub-surface water and reasonable locations for sampling the Water Ice and Subsurface Deposit Observation On Mars ground-penetrating radar is an instrument to characterise the stratigraphy under the surface together with Adron, which can provide informations on sub-surface water.

Adron

The Adron instrument searches sub-surface water and hydrated minerals and will be used in combination with the WISDOM instrument to investigate the ground structure under the surface to search for areas for drilling and sample collection.

Mars Multispectral Imager for Subsurface Studies (Ma_MISS)

The Ma_MISS instrument is located inside the drill. It will help to investigate the mineralogy and rock formation on the Mars.

MicrOmega

The MicrOmega instrument is an imaging spectrmeter in visible and infrared range. It is used for mineralogy investigations on Martian samples.

Raman Spectrometer (RLS)

With the help of the Raman spectrometry it is possible to establish the mineralogical composition of samples and to identy organic pigments.

Mars Organic Molecule Analyser (MOMA)

The Mars organic molecule analyser will detect biomarkers in ground samples. These biomarkers will help to answer questions concerning the potential origin and distribution of life on Mars.

13.6.2 Mars Sample Return Mission

There are several reasons for an exploration of Mars because it is the most Earth-like planet in our Solar System. Geological structures show that liquid water flowed on the surface of Mars long ago and Mars is the most accessible planet for evaluating whether or not life exists elsewhere in the Universe, or has existed.

The Mars Sample Return (MSR) mission is a challenging and complex mission and it could be launched between 2020 and 2022. New technologies will be required. The mission calls for five spacecrafts. An Earth re-entry vehicle including the landing system on Mars, the Mars ascent/descent vehicle, the rendezvous system in the Mars orbit and an Earth re-entry vehicle. Furthermore, careful measures will be needed to protect the samples to avoid contamination of Mars by organisms from Earth and *vice versa*. A Mars sample return mission would need to comply with planetary protection requirements on Earth and on Mars as well.

13.7 Estimating Masses and Costs

Due to restricted technical means (presently up to $20\,000\,\text{m}\cdot\text{s}^{-1}$ of velocity requirement, Apollo: approximately $17\,000\,\text{m}\cdot\text{s}^{-1}$, significantly) options C and D can be discarded from the outset for an implementation with the today's means. Option B should be further examined here, because a flight time of up to three years in option A takes a long time with all the risks and disadvantages involved. In the following a rough estimate of the masses and costs is carried out for this purpose. Therefore, let us assume an international crew of three spacemen (one astronaut, one cosmonaut, and one taikonaut or European).

Estimating masses of option B

We enlarge (approximately 5.5 t) the flight unit concerned of the Apollo command module assuming a net mass of 10 t. Supply goods of 4.5 t are added to this for return flight of 300 days and 4 t for outward flight of 270 days. Three flight-tested technologies are available as energy source.

- Solar cells
- Fuel cells
- Radioisotope thermoelectric generators (RTGs).

An installed electrical output of 20 kW is planned, so that optional electric engines can be used supportingly. To provide this power by solar cells, an area of approximately $100\,\text{m}^2$ is required. Close to Mars this power is reduced to below 10 kW by the influence of the square of the distance from the Sun while at the perihel P_4 the power exceeds 90 kW.

Fuel cells are not subjected to these instabilities. With a mean energy density of $700\,\text{Wh}\cdot\text{kg}^{-1}$ an installed capacity of approximately $19\,\text{t}\cdot\text{kW}^{-1}$ for energy supply is required for 550 days. Radioisotope thermoelectric generators have a higher energy density. They contain plutonium-238 with a half-life of 89 years. A thermal energy of 390 W per kilogramme of plutonium is released which is converted into electric power with an efficiency of approximately 10 %. For 20 kW of thermal energy, RTGs with a mass of approximately 1 t are required.

The lander must not exceed a net mass of 2 t for two spacemen. A mass of 1 t needs to be added to the landing system consisting of parachutes and a heat shield. Residence on the Martian surface should not exceed one week and the rock samples collected should not have a total mass of more than 100 kg. Energy supply is ensured by RTGs. For departure from the Martian orbit and return flight a velocity of $4\,460\,\text{m}\cdot\text{s}^{-2}$ is required. Based on the assumption of using storable propellants, monomethylhydrazine (MMH) as fuel and dinitrogen tetroxide (NTO) as oxidiser, with an exhaust velocity of $3\,200\,\text{m}\cdot\text{s}^{-1}$ of the combustion gases, a propellant mass of approximately 13 t results. The total mass of the lander amounts to 18 t.

A service module is available for landing and return flight. The unit for landing is hereinafter referred to as 1st stage. The unit for return flight is referred to as 2nd stage and the unit for the perihel manoeuvre P_4 is referred to as 3rd stage. All of the three stages should propelled by a cryogenic propellant combination consisting of liquid hydrogen (LH2) as fuel and liquid oxygen (LOX) as oxidiser. The exhaust velocity of the combustion gases, such as water vapour and hydrogen, should be $4500 \, m \cdot s^{-1}$. The 1st stage performs manoeuvres at E_1 and at M_2, and is then separated and left in the Martian orbit, while the 2nd stage initiates return flight at M_3. The 3rd stage performs the manoeuvre required at P_4. The 2nd stage and 3rd stage are additionally equipped with electric engines to shorten flight time as well as to head for planet Venus for swing-by manoeuvre without landing. Xenon with an exhaust velocity of $50000 \, m \cdot s^{-1}$ should be used as propellant. The installed power of the electric engines should be 10 kW which could generate a permanent thrust of 0.4 N. The resulting low acceleration would hardly be noticed by the spacemen, but only slight disturbances of weightlessness. As a result of a swing-by manoeuvre at Venus (at X_1 or X_2) a maximum velocity of $8870 \, m \cdot s^{-1}$ could be used theoretically. In our case, only $5000 \, m \cdot s^{-1}$ should be available in practice.

Now, the 3rd stage of our spacecraft can be estimated. A propellant mass of 16 t is required for a payload of 15 t (cabin and supplies) and for manoeuvre at P_4 with $3000 \, m \cdot s^{-1}$. During approach the total mass amounts to 34 t at P_4.

So, the 2nd stage of our spacecraft can be estimated. A propellant mass of 146 t is required for a payload of 40 t (3rd stage and additional supplies) and for manoeuvre at M_3 (with $5500 \, m \cdot s^{-1}$). The total mass amounts to 206 t before departure from M_3. The mass of the 2nd stage (206 t) together the mass of the lander (18 t) results in a payload of 230 t for the 1st stage including additional supplies. This results again in an estimated mass of 837 t of propellant for the 1st stage which performs the manoeuvres at E_1 and M_2 with $5500 \, m \cdot s^{-1}$ at a total mass of 1187 t at departure position E_1.

The use of MMH/NTO as propellant in the two propulsion stages is also possible, but requires a considerably higher mass of propellant. Therefore, two stages are necessary for manoeuvres at E_1, M_2, M_3, and at P_4, which will then be separated each. Storage of cryogenic LH2/LOX as propellant requires special isolation measures which have not (yet) been flight proven. Basically, these techniques are possible and are already used today in case of liquid helium with a storage life of up to 5 years.

In addition to the possibilities to shorten flight time mentioned above, two other possibilities should be described briefly: swing-by at the Moon and the use of solar sailing. Theoretically, the Moon has a maximum swing-by requirement of $1679 \, m \cdot s^{-1}$ according to its near-surface orbital velocity. At best, $1500 \, m \cdot s^{-1}$ can be utilised thereof, which is of subordinate importance for our mission planning. Since a defined position of the Moon is required for this gravitational manoeuvre, mission planning will be further limited thereby and the use seems rather unlikely. Also the use of a solar sail seems rather unlikely. The sunlight generates a thrust of approximtely $8\,N \cdot km^{-2}$ on a completely mirrored, tearproof sail area in a near-earth distance from the Sun. The mass of the sail must not be more than $5\,g \cdot m^{-2}$ (for comparison: normal paper weighs $80\,g \cdot m^{-2}$!). At the perihel of the returning orbit the thrust increases fourfold according to the square of the distance from the Sun and thus could contribute to a shortening of flight time substantially. The technique of unfolding of a solar sail is very complicated and a reliable solution of this problem is not (yet) in sight. Therefore, it should not be considered in this elaboration.

Estimating costs of option B

From roughly estimated masses a similar rough estimate of the costs involved can be derived. From the Earth to near-earth orbit one kilogramme of payload costs approximately 10 000 euros in transportation costs, which totals to about $1187\,t \cdot 10$ million euros $= 12$ billion euros. Around 40 launchers such as Ariane 5 or a similar number of flights of the Space Shuttle are required. In contrast, an unmanned MSRO mission with an option A trajectory requires only one flight of an Ariane 5 launcher (maximum of two flights, if the Mars lander and the propulsion units have to be launched separately). One kilogramme of flight hardware (without propellant) costs approximately 100 000 euros. A reserve unit is required for all eventualities.

Development models and qualification models require three times the amount of one flight unit. As a result, we get a sum of 88 billion euros for a net mass of approximately 175 t. Thus, the total costs of 100 billion euros are in the same order of magnitude as the costs for the International Space Station (ISS), the most expensive space project so far. The low profitability with comparatively low technical and human risks of the International Space Station is subject of increasing criticism among more and more persons responsible.

Estimating masses of option A

If the problem of long-term storage of cryogenic propellants cannot be solved (and there are many indications) there remains only the possibility of using storable propellants (MMH/NTO) and thus, option A as possible mission. This requires a 'waiting' for the next opposition and a remaining in the sphere affected by Mars of approximately 15 months. Therefore, it makes sense to provide another habitation module of 20 t for supplies sufficient for a few days on the Martian surface, and a Mars vehicle to increase mobility in addition to the lander as already mentioned before. This module allows to stay on our neighbouring planet of more than one year and it will be left on the surface later on. The landing is carried out together with the launch stage (then we have got a two-staged lander) or separately with propellant saving, direct injection heading to the Martian surface.

So, three propulsion stages of the spacecraft can be estimated again. We each require a stage mass ratio of $(m_1/m_2) = 8$. A propellant mass of 16 t is required for a payload of 15 t for the crew cabin and supplies, and a departure manoeuvre at M_3 with 2100 m·s^{-1}. The total mass before departure at M_3 amounts to 34 t.

So, the 2nd stage of the spacecraft can be estimated again. A propellant mass of 43 t is required for a payload of 40 t (3rd stage and additional supplies) and departure manoeuvre at M_2 with 2100 m·s^{-1}. The total mass before departure at M_2 amounts to 89 t. The two landers (38 t) have been separated before and were flown directly to the Martian surface. Without separation the propellant consumption for this manoeuvre would roughly double.

The 2nd stage (89 t) together with the landers (38 t) results in a payload of 130 t for the 1st stage including additional supplies. This results in an estimated propellant mass of 385 t for the entire Mars spacecraft at E_1 with 3600 m·s^{-1} at a total mass of 570 t before departure at E_1.

Estimating costs of option A

The costs for this option only insignificantly differ from the costs of option B for the first approximation, since the spacecrafts are not in essence different among themselves. The transportation costs only halve during transport of the components from Earth into orbit, since only half the flights of Ariane 5 launchers or Space Shuttles are required. Since the duration of stay in the sphere affected by Mars takes much longer than in option B, considerable additional financial expenditures have to be planned for this. Previously, the American space agency (NASA) already estimated the costs at 400 billion US dollars for a manned flight to Mars for this option.

13.8 Conclusions

The results presented show that it is possible in principle to take man from Earth to Mars and back to Earth. It is estimated that there will be no Martians in the 21th century due to high technical and financial risks. On the other hand, this means that no Earthling living today will ever experience a flight to Mars as an observer. Financial necessities and risks for men make it less likely that there will be manned missons to Mars and economic exploitations of the planet for the foreseeable future. The realisation, if anything, would require similar financial efforts as of the International Space Station, (ISS), which can only be overcome by international cooperation. The expected knowledge gained is low and the risk to humans is high. In future, unmanned missions like NASA's Mars Sample Return Orbiter (MSRO) will be therefore carried out as in option A. This option, and also option B, was already described for the first time by *Walter Hohmann*[21] in 1925. Regular operations between Earth and Mars, in option C and D, require a complete new propulsion technology. From our present point of view this technology can only be found in the utopian realms of science fiction.

13.9 Questions for Further Studies

1.   What factors hinder manned space flight and colonisation of outer space from rapid development as has hoped by early enthusiasts?
2.   Give boundary conditions of interstellar space flight.
3.   What kind of interstellar signal is expected to be found by the SETI project?
4.   Do you think that the SETI project will present evidence for extraterrestrial life?
5.   Give reasons why to go back to the Moon, on to Mars and beyond.
6.   What are the emotional and psychological impacts of a Mars mission with no return?
7.   What are the names of the two scientific Mars rovers?
8.   Give the names of the three groups of microorganisms the two Mars rovers are searching for.
9.   At what interval minimum-energy launch windows occur for a mission to Mars?
10. What are the key environmental factors of a Martian terraforming that need to be overcome?

---

21 Hohmann W. Die Erreichbarkeit der Himmelskörper, Oldenbourg Verlag, München, 1925.

14 Appendix

14.1 Acronyms and Abbreviations

A

a	Year (lat. annum)
A	Unit of electric current (Ampere)
A5	Ariane 5
A62	Arine 6 configuration with two solid rocket boosters
A64	Arine 6 configuration with four solid rocket boosters
ABC	German collective term for biological, chemical, and nuclear
ABM	Anti-ballistic missile
ABS	Absolute value
ACS	Arc cosine
AU	Astronomical unit
AFB	Air force base
AIAA	American Institute of Aeronautics and Astronautics
Airbus DS	Airbus Defense and Space Corporation
AIT	Assembly, integration, testing
Al	Chemical symbol for aluminium
AOCS	Attitude and orbit control system
APCP	Ammonium perchlorate composite propellant
APU	Auxiliary power unit
ARD	Atmospheric re-entry demonstrator
ASN	Arc sine
ATN	Arc tangent
ATV	Automated transfer vehicle
Au	Chemical symbol for gold (lat. aurum)

B

BDLI	German Aerospace Industries Association
BOL	Begin of life
BT	Bilan technique

C

C	Chemical symbol for carbon
C/SiC	Carbon/Silicon carbide
C-C/SiC	Carbon – Carbon/Silicon carbide
CAD	Computer aided design
CAM	Computer aided manufacturing

C

CC	Combustion chamber
CCB	Change control board
CD	Discharge coefficient
Cd	Unit of luminous intensity (Candela)
CDR	Critical design review
CE	Efficient exhaust velocity
CEO	Chief executive officer
CEST	Central European summer time
CET	Central European time
CF	Cold flow
CFRP	Carbon fibre reinforced plastics
CHT	Catalytic hydrazine thruster
CIA	Central Intelligence Agency
CM	Configuration management
Co	Chemical symbol for cobalt
CPM	Chemical propulsion module
Cr	Chemical symbol for chrome
CRE	Commision revue d'essais
CRV	Crew rescue vehicle
Cs	Chemical symbol for caesium
CSG	Centre Spatial Guyanais
Cu	Chemical symbol for copper

D

DGLR	German Society for Aerospace Technology
DIN	German term for German industrial standard
DLR	Deutsches Zentrum für Luft- und Raumfahrt e.V.
DM	Development model
DP	Data processing
DRB	Delivery review board

E

EADS	European Aeronautic Defence and Space Company Now: Airbus Group
EAP	L'Etage d'accélération à poudre
EB	Electron beam
EDP	Electronic data processing
EITA	Electron bombardment ion thruster assembly
ELDO	European Launcher Development Organisation

E

EOL	End of life
EOR	Electrical orbit raising
EPC	L'Etage principal cryotechnique
EPC	L'Etage principal cryotechnique
EPS	L'Etage propulsif à ergols stockables
EQSR	Equipment qualification & suitability (status) review
ESRO	European Space Research Organisation
ETR	Eastern test range
EVA	Extra vehicular activity
EXP	Exponential function

F

F	Unit for electrical capacitance (Farad)
F	Chemical symbol for fluorine
FC	Fuel cell
Fe	Chemical symbol for iron (lat. ferrum)
FFE	German term for independent research and development
FGSE	Fluid ground support equipment
FM	Flight model
Fu	Fuel

G

GATT	General Agreement on Tariffs and Trade
GEO	Geostationary orbit
GFRP	Glassfiber re-inforced plastics
GH2	Gaseous hydrogen
GHe	Gaseous helium
GMES	Global monitoring for environment and security
GMT	Greenwich mean time
GN2	Gaseous nitrogen
GNSS	Global navigation satellite system
GPS	Global positioning system
GSE	Ground support equipment
GTO	Geo transfer orbit

H

h	Hour
H	Chemical symbol for hydrogen
H	Unit of inductance (symbol H)
HCU	Hard coal unit

H

He	Chemical symbol for helium
HF	High frequency
Hg	Chemical symbol for mercury
HM60	Haute Moteur 60 (Ariane 5 main propulsion engine)
HP	High pressure
HTPB	Hydroxyl-terminated polybutadiene
HT	High temperature
Hz	Unit of frequency (Hertz)

I

IAF	International Astronautical Federation
IBIT	Impulse bit
ICBM	Intercontinental ballistic missile
In	Inch
Ir	Chemical symbol for iridium
IR	Infrared
IRBM	Intermediate range ballistic missile
IRR	Integration readiness review
ISO	International Standard Organisation
ISO	Infrared Space Observatory
ISP	Specific impulse
ISS	International Space Station
IT	Information technology
ITU	International Telecommunication Union

J

J	Unit of energy (Joule)

K

K	Chemical symbol for potassium
K	Unit of the absolute temperature (Kelvin)
kg	Kilogramme
KIP	Key inspection point

L

L1 – L5	Lagrangian points
L9	Ariane 5 upper stage
LED	Light emitting diode
LEO	Low earth orbit

L

LH2	Liquid hydrogen
LHe	Liquid helium
LJ	Light year
LJ	Light year
LN	Natural logarithm
LN2	Liquefied nitrogen
LNG	Liquefied natural gas
LOG	Logarithm
LOX	Liquefied oxygen
LP	Low pressure

M

m	Metre
Ma	Mach number
MAIT	Manufacturing, assembly, integration, test
MAK	German term for maximum allowable concentration
MEO	Middle Earth orbit
MEOP	Maximum expected operating pressure
Mg	Chemical symbol for magnesium
MGSE	Mechanical ground support equipment
MIP	Mandatory inspection point
MMH	Monomethylhydrazine
MMU	Manned manoeuvring unit
Mn	Chemical symbol for manganese
Mo	Chemical symbol for molybdenum
MPDT	Magnetoplasmadynamic thruster
MRB	Material review board
MRR	Manufacturing readiness review

N

N	Chemical symbol for nitrogen
N	Unit of force (Newton)
NATO	North Atlantic Treaty Organisation
NDI	Non destructive inspection
Ni	Chemical symbol for nickel
NMD	National Missile Defense
NN	Mean sea level
NSSK	North-south station keeping
NTO	Dinitrogen tetroxide

O

O	Chemical symbol for oxygen
OMS	Orbital manoeuvring system
OPEC	Organisation of Petroleum Exporting Countries
Ox	Oxidiser

P

P	Chemical symbol for phosphor
PA	Product assurance
PAR	Programm appraisal review
Pb	Chemical symbol for lead
PBAN	Polybutadiene acrylonitrile
PCA	Pressure control assembly
PDM	Pre-development model
PDR	Pre-design review
PFE	Parallel flow equipment
PIA	Propellant isolation assembly
PFE	Parallel flow equipment
PIA	Propellant isolation assembly
PM	Project management
PQM	Pre-qualification model
PSD	Pogo suppression device
Pt	Chemical symbol for platinum
PTFE	Polytetrafluoroethylene
Pu	Chemical symbol for plutonium

Q

QA	Quality assurance
qm	Square metre
QM	Qualification model
QTRR	Qualification test readiness review

R

RADAR	Radio detection and ranging
RAMS	Reliability, availability, maintenance, safety
RCS	Reaction control system
Re	Chemical symbol for rhenium
Rh	Chemical symbol for rhodium
RITA	Radiofrequency ion thruster assembly
ROI	Return on investment

R

RP	Rocket propellant
RP	Rapid prototyping
rpm	Rotations per minute
Ru	Chemical symbol for ruthenium

S

s	Second
S	Chemical symbol for sulphur
SALT	Strategic arms limitation talks
SCA	Système contrôle d'attitude
scc	Standard cubic centimetre
SDI	Strategic Defense Initiative
SETI	Search for extraterrestrial intelligence
Si	Chemical symbol for silicon
SL	Sea level
SLBM	Submarine launched ballistic missile
SRB	Solid rocket booster
SS	Subsystem
SSME	Space shuttle main engine
SSO	Sun synchronous orbit
STS	Space transportation system

T

T	Unit of magnetic flux density (Tesla)
TC	Thrust chamber
TCA	Thrust chamber assembly
TGSE	Tanking ground support equipment
Ti	Chemical symbol for titanium
TIG	Tungsten inert gas
TRR	Test readiness review

U

U	Chemical symbol for uranium
UDMH	Unsymmetrical dimethylhydrazine
UNO	United Nations Organisation
UPS	Unified propulsion system
USAF	US Air Force
UV	Ultraviolet

V

V	Chemical symbol for vanadium
VDI	Association of German Engineers
VEB	Vehicle electronics bay
VEEGA	Venus-Earth-Earth-Gravity Assist
VTOL	Vertical takeoff and landing

W

W	Chemical symbol for tungsten (also known as wolfram)
W	Watt
WIG	Tungsten inert gas
WTR	Western test range
WWW	World wide web

X

Xe	Chemical symbol for xenon
XMM	X-ray multi-mirror mission

14.2 Prefixes and Quantities

Table 14.1: Prefixes for indication of decimal multiples and fractions of units.

Prefix name	Decimal power	Symbol
Yotta	10^{24}	Y
Zetta	10^{21}	Z
Exa	10^{18}	E
Peta	10^{15}	P
Tera	10^{12}	T
Giga	10^{9}	G
Mega	10^{6}	M
Kilo	$10^{3} = 1,000$	k
Hecto	$10^{2} = 100$	h
Deca	$10^{1} = 10$	da
	$10^{0} = 1$	
Deci	$10^{-1} = 0.1$	d
Centi	$10^{-2} = 0.01$	c
Milli	$10^{-3} = 0.001$	m
Micro	10^{-6}	µ
Nano	10^{-9}	n
Pico	10^{-12}	p
Femto	10^{-15}	f
Atto	10^{-18}	a
Zepto	10^{-21}	z
Yocto	10^{-24}	y

Table 14.2: Relevant astronautical variables.

Symbol	Quantity	Symbol	Quantity
a	Major semi-axis	t	Time
b	Minor semi-axis	T	Orbital period
e	Eccentricity	v	Velocity
G	Gravitational constant	x, y	Cartesian coordinates
g	Gravitational acceleration	z	Ceiling, maximum altitude
i	Inclination (=slope)	β	Orbital slope angle
l	Tether length	χ	Ballistic path angle
M	Mass of celestial body	φ	Apse angle (true anomaly)
n	Unaffected flyby distance	Φ	Deflection angle
N	Payload	λ	Ballistic factor
p	Apse vector	μ	Standard gravity parameter
$\bar{r}$	Position vector	ψ	Position tangent angle
R	Radius of a celestial body	ω	Argument of periapsis[22]
s	Range, throw distance	Ω	Tight ascension

Table14.3: Mathematical symbols and subscripts.

Symbol	Quantity	Subscript	Name
acs	Arc cosine	0	Initial
atn	Arc tangent	a	Exhaust
cos	Cosine	apo	Apoapsis[23]
e	Exponential function	ball	Ballistic
	to the base of 2,718282...	CG	Centre of gravity
ln	Natural logarithm	CM	Centre of mass
sin	Sine	e	End
tan	Tangent	E	Earth
π	Ludolph's constant	orbit	Orbital
	3,1415926535...	max	Maximum
		M	Metacentre
		n	Normal
		loc	Local
		peri	Periapsis
		r	Radial
		syn	Synodic

---

22 The periapsis is the closest point of an orbit from the centre of a celestial body.
23 The apoapsis is the most distant point of an orbit from the centre of a celestial body.

14.3 Formulary of Classical Orbital Mechanics

Geometry

Polar equation:

$$r = \frac{p}{1 + e\cos\varphi}$$

(14.1)

Ellipse ($e < 1$):

$$\frac{x^2}{a^2} + \frac{y^2}{b^2} = 1$$

(14.2)

$$x = a\cos\varphi$$

(14.3)

$$y = b\sin\varphi$$

(14.4)

$$a = \frac{p^2}{p} = \frac{p}{1-e^2} = \frac{b}{\sqrt{1-e^2}} = \frac{r_{peri} + r_{apo}}{2}$$

(14.5)

$$b = \sqrt{ap} = a\sqrt{1-e^2} = \frac{p}{\sqrt{1-e^2}}$$

(14.6)

$$p = \frac{b^2}{a} = a\left(1-e^2\right) = b\sqrt{1-e^2}$$

(14.7)

$$e = \sqrt{1 - \frac{b^2}{a^2}} = \sqrt{1 - \frac{p}{a}} = \sqrt{1 - \frac{p^2}{b^2}}$$

(14.8)

$$r_{peri} = a\left(1-e\right) = \frac{p}{1+e}$$

(14.9)

$$r_{apo} = a\left(1+e\right) = \frac{p}{1+e}$$

(14.10)

Parabola ($e = 1$):

$$r = \frac{p}{1 + \cos\varphi}$$

(14.11)

$$r = \frac{p}{1 + \cos\varphi}$$

(14.12)

$$r_{peri} = \frac{p}{2}$$

(14.13)

Hyperbola ($e > 1$):

$$\frac{x^2}{a^2} - \frac{y^2}{b^2} = 1 \tag{14.14}$$

$$x = a\sin\varphi \tag{14.15}$$

$$y = b\sin\varphi \tag{14.16}$$

$$\varphi_{max} = \mathrm{acs}\left(-\frac{1}{e}\right) \tag{14.17}$$

$$a = \frac{b^2}{p} = \frac{p}{e^2-1} = \frac{b}{\sqrt{e^2-1}} \tag{14.18}$$

$$b = \sqrt{a \cdot p} = a\sqrt{e^2-1} = \frac{p}{\sqrt{e^2-1}} \tag{14.19}$$

$$p = \frac{b^2}{a} = a\left(e^2-1\right) = b\sqrt{e^2-1} \tag{14.20}$$

$$e = \sqrt{q + \frac{b^2}{a^2}} = \sqrt{1 + \frac{p}{a}} = \sqrt{1 + \frac{p^2}{b^2}} \tag{14.21}$$

$$r_{peri} = a\left(e-1\right) = \frac{p}{1+e} \tag{14.22}$$

Orbit determination

Standard gravitational parameter:

$$\mu = G \cdot M \tag{14.23}$$

Vis viva equation:

$$v = \sqrt{\mu\left(\frac{2}{r} - \frac{1}{a}\right)} \quad \Longleftrightarrow \quad a = \left[\frac{2}{r} - \frac{v^2}{\mu}\right]^{-1} \tag{14.24}$$

Eccentricity:

$$e = \sqrt{1 + \left(\frac{r \cdot v^2}{\mu} - 2\right)\frac{r \cdot v^2\sin^2\Psi}{\mu}} = \left|\frac{1}{\cos\varphi + \tan\varphi \cdot \sin\varphi}\right| \tag{14.25}$$

Position tangent angle:

$$\Psi = 180° - \text{atn}\left[\frac{1+e\cos\varphi}{e\sin\varphi}\right] \qquad (14.26)$$

Rotational velocity:

$$v_r = \frac{r \cdot v \sin\Psi}{a(1-e^2)} e\sin\varphi \qquad (14.27)$$

Normal velocity:

$$v_n = \frac{r \cdot v \sin\Psi}{a(1-e^2)}(1+e\cos\varphi) \qquad (14.28)$$

Angular velocity:

$$\dot{\varphi} = \frac{r \cdot v \sin\Psi}{a^2(1-e^2)^2}(1+e\cos\varphi)^2 \qquad (14.29)$$

Orbital period:

$$T = 2\pi\sqrt{\frac{a^3}{\mu}} \qquad (14.30)$$

Time period of elliptic trajectories:

$$t = \sqrt{\frac{a^3}{\mu}}\left[2\,\text{atn}\left(\sqrt{\frac{1-e}{1+e}}\tan\frac{\varphi}{2}\right) - \frac{e\sqrt{1-e^2}\sin\varphi}{1+e\cos\varphi}\right] \qquad (14.31)$$

Time period of parabolic trajectories:

$$t = \sqrt{\frac{p^3}{\mu}}\left[\frac{1}{2}\tan\frac{\varphi}{2} + \frac{1}{6}\tan^3\frac{\varphi}{2}\right] \qquad (14.32)$$

Time period of hyperbolic trajectories:

$$t = \sqrt{\frac{p^3}{\mu}}\frac{1}{e^2-1}\left[\frac{e\sin\varphi}{1+e\cos\varphi} - \frac{1}{\sqrt{e^2-1}}\ln\left|\frac{(e-1)\tan\frac{\varphi}{2}+\sqrt{e^2-1}}{(e-1)\tan\frac{\varphi}{2}-\sqrt{e^2-1}}\right|\right] \qquad (14.33)$$

Acceleration of gravity:

$$g = \frac{\mu}{r^2} \tag{14.34}$$

Residual velocity in infinity:

$$v_\infty = \sqrt{v^2 - \frac{2\mu}{r}} \tag{14.35}$$

Orbital velocity requirement:

$$\Delta v_{circle} = \sqrt{v_{circle}^2 + v_{loc}^2 - 2 v_{circle} v_{loc} \sin \Psi} \tag{14.36}$$

Synodic period:

$$t_{syn} = \left| \frac{1}{\dfrac{1}{T_1} - \dfrac{1}{T_2}} \right| \tag{14.37}$$

Swing-by

Unaffected swing-by distance:

$$n = r_{peri} \sqrt{1 + \frac{2\mu}{r_{peri}\, v_\infty^2}} \tag{14.38}$$

Deflection angle:

$$\Phi = 180° - 2 \operatorname{atn} \frac{n \cdot v_\infty}{\mu} \tag{14.39}$$

Orbital transitions

Hohmann transition:

$$\Delta v = \Delta v_1 + \Delta v_2 \tag{14.40}$$

with

$$\Delta v_1 = \sqrt{\frac{\mu}{r_1}} \left(\sqrt{\frac{2 r_2}{r_1 + r_2}} - 1 \right) \tag{14.41}$$

and

$$\Delta v_2 = \sqrt{\frac{\mu}{r_2}} \left(1 - \sqrt{\frac{2r_1}{r_1 + r_2}} \right)$$
(14.42)

3-impulse transfer:

$$\Delta v = \left(\sqrt{2} - 1 \right) \left(\sqrt{\frac{\mu}{r_1}} + \sqrt{\frac{\mu}{r_2}} \right)$$
(14.43)

Spiral transfer:

$$\Delta v = \sqrt{\frac{\mu}{r_1}} \left(1 - \sqrt{\frac{r_1}{r_2}} \right)$$
(14.44)

Ballistics

Ballistics factor:

$$\lambda = \frac{R_0 \cdot v_0^2}{\mu}$$
(14.45)

Eccentricity:

$$e = \sqrt{1 - \lambda \left(2 - \lambda \right) \cos^2 \beta}$$
(14.46)

Ceiling (maximum altitude):

$$z = R_0 \left(\frac{e + \lambda - 1}{2 - \lambda} \right)$$
(14.47)

Range:

$$s = 2R_0 \, \text{acs} \left(\frac{1 - \lambda \cos^2 \beta}{e} \right)$$
(14.48)

Maximum range:

$$s_{\text{max}} = 2R_0 \, \text{acs} \left(\frac{2\sqrt{1 - \lambda}}{2 - \lambda} \right)$$
(14.49)

Orbital inclination angle for maximum range:

$$\beta_{\text{max}} = \text{acs} \left(\frac{1}{\sqrt{2 - \lambda}} \right)$$
(14.50)

Flight time:

$$t_{ball} = \frac{2\mu}{v_0^3}\left(\frac{\lambda}{2-\lambda}\right)^{1.5}\left(\chi + e\sin\chi\right) \tag{14.51}$$

with

$$\chi = \text{acs}\left(\frac{1-\lambda}{e}\right) \tag{14.52}$$

Influences of Earth's oblateness

Shifting of right ascension (degrees per day):

$$\dot{\Omega} = \frac{-9.96\cos i}{\left(\dfrac{a}{R_E}\right)^{3.5}\left(1-e^2\right)^2} \tag{14.53}$$

Shifting of the argument of perigee (degrees per day):

$$\dot{\omega} = \frac{4.98\left(5\cos^2 i - 1\right)}{\left(\dfrac{a}{R_E}\right)^{3.5}\left(1-e^2\right)^2} \tag{14.54}$$

Tethers

Centre of mass:

$$r_{CM} = \frac{m_1 \cdot r_1 + m_2 \cdot r_2}{m_1 + m_2} \tag{14.55}$$

Centre of gravity:

$$r_{CG} = \sqrt{\frac{m_1 + m_2}{\dfrac{m_1}{r_1^2} + \dfrac{m_2}{r_2^2}}} \tag{14.56}$$

Metacentre:

$$r_{MC} = \sqrt[3]{\frac{m_1 \cdot r_1 + m_2 \cdot r_2}{\dfrac{m_1}{r_1^2} + \dfrac{m_2}{r_2^2}}} \tag{14.57}$$

with

$$r_{MC}^3 = r_{CM} \cdot r_{CG}^2 \qquad (14.58)$$

Orbital period:

$$T = 2\pi \sqrt{\frac{r_{MC}^3}{\mu}} \qquad (14.59)$$

Final acceleration of tethered systems:

$$g = \mu \left[\frac{1}{\left(r_{MC} + l\right)^2} - \frac{r_{MC} + l}{r_{MC}^3} \right] \qquad (14.60)$$

14.4 Formulary of Rocket Flights

Tsiolkovski rocket equation:

$$v_e = v_a \ln \frac{m_0}{m_e} \qquad (14.61)$$

Payload without return flight:

$$N = \frac{e^{\frac{v_e}{v_a}} m_e - m_0}{1 - e^{\frac{v_e}{v_a}}} \qquad (14.62)$$

Payload of an empty return flight:

$$N = \frac{e^{\frac{2 v_e}{v_a}} m_e - m_0}{1 - e^{\frac{v_e}{v_a}}} \qquad (14.63)$$

Payload of an empty outward flight:

$$N = \frac{e^{v_e \cdot v_a^{-1}} m_e - \dfrac{m_0}{e^{v_e \cdot v_a^{-1}}}}{1 - e^{v_e \cdot v_a^{-1}}} \qquad (14.64)$$

Payload without empty flight:

$$N = \frac{e^{\frac{2 v_e}{v_a}} m_e - m_0}{1 - e^{\frac{2 v_e}{v_a}}} \qquad (14.65)$$

14.5 Websites

www.aiaa.org
American Institute of Aeronautics and Astronautics. Accessible in English.

www.arianespace.com
Home page of Arianespace company. Accessible in English and French.

www.astroinfo.org
Astronomical informations. Accessible in German.

www.astronautix.com
Encyclopaedia Astronautica by Mark Wade. Accessible in English.

www.astronews.com
Current press release. Accessible in German.

www.boeing.com
Home page of Boeing company. Accessible in English.

www.calsky.com
Computer programmes for astronomy and orbital mechanics. Accessible in German and English.

www.cnes.fr
French Space Agency. Accessible in French and English

www.cnsa.gov.cn
China National Space Administration. Accessible in English and Chinese.

www.dlr.de
Home page of Deutsches Zentrum für Luft- und Raumfahrt e.V. Accessible in German and English.

www.airbusgroup.com
Home page of Airbus (former: EADS). Accessible in German, English, French, Spanish, and Chinese.

www.esa.int
European Space Agency. Accessible in German, English, French, Czech, Danish, Dutch, Finnish, Greek, Italian, Norwegian, Polish, Portuguese, Romanian, and Swedish.

www.federalspace.ru
Russian Space Agency ROSKOSMOS. Accessible in Russian and English.

www.isro.gov.in
Indian Space Research Organisation. Accessible in English and Hindi.

global.jaxa.jp
Japan Aerospace Exploration Agency. Accessible in English.

www.heavens-above.com
Continuously updated data of satellites and sky observations. Accessible in English.

www.jpl.nasa.gov
Home page of NASA Jet Propulsion Laboratory at Caltech. Scientific space missions, with the participation of NASA. Accessible in English.

www.ksc.nasa.gov
Home page of NASA Kennedy Space Center. Accessible in English.

www.n2yo.com
Real time satellite tracking and predictions. Accessible in English.

www.nasa.gov
Home page of the National Aeronautics and Space Administration. Accessible in English.

www.raumfahrer.net
Website of the German non-profit Raumfahrer Net e. V., Regensburg. Accessible in German.

www.russianspaceweb.com
News and history of astronautics in the former USSR. Accessible in English.

space.skyrocket.de
Gunter's space page. Accessible in English.

www.space.com
Current press release. Accessible in English.

www.spaceflight101.com
Covers current space flight events with special focus on the technical aspects of space operations with long range launch tracker. Accessible in English.

www.spaceflightnow.com
Source for space news. Accessible in English.

www.spacenews.com
Covers the business and politics of the global space industry. Accessible in English.

www.stk.com
Analysis software for land, sea, air, and space. Accessible in English.

14.6 Credits for Illustrations

[1] http://images.ksc.nasa.gov/photos/1969/captions/AS11-40-5903.html
(retrieved 01.01.2015)

[2] Orbital Debris. Quarterly News. Volume 18, Issue 1, January 2014.
http://orbitaldebris.jsc.nasa.gov/newsletter/newsletter.html (retrieved 01.02.2015)

[3] COSPAR Committee for the COSPAR International Reference Atmosphere. CIRA 1972:
COSPAR International Reference Atmosphere 1972. Akademie Verlag, Berlin, 1972.

[4] http://eoimages.gsfc.nasa.gov/images/imagerecords/57000/57752/
land_shallow_topo_2048.jpg (retrieved 02.01.2015).

[5] http://spaceflight.nasa.gov/gallery/images/shuttle/sts-79/html/s79e5327.html
(retrieved 16.07.2015)

[6] Hall R (Ed). The History of Mir. 1986 – 2000. British Interplanetary Society, London, 2000
Hall R (Ed). Mir. The Final Year. British Interplanetary Society, London, 2001

[7] http://science.nasa.gov/science-news/science-at-nasa/2002/06nov_ssme/
(retrieved 18.12.2015)

[8] Lainé R. Ariane. The European Launcher. European Space Agency. Paris, 2005
Ariane KTH 2005.pdf (retrieved 06.05.2015)

[9] http://grin.hq.nasa.gov/abstracts/GPN-2000-001156.html (retrieved 13.04.2015)

[10] http://www.nasa.gov/sites/default/files/thumbnails/image/ksc-69pc-442.jpg
(retrieved 16.04.2015)

[11] http://www.esa.int/spaceinimages/Images/2016/04/Artist_s_view_of_the_configuration_of_
Ariane_6_using_four_boosters_A643 (retrieved 12.04.2016)

[12] http://www.dfrc.nasa.gov/Gallery/Photo/ALT/Medium/ECN-6887.jpg
(retrieved 04.10.2015)

[13] http://science.ksc.nasa.gov/shuttle/missions/sts-99/images/high/KSC-00PP-0223.jpg
(retrieved 27.08.2015)

[14] http://www.nasa.gov/sites/default/files/thumbnails/image/sls_block1_foam_afterburner.jpg
(retrieved 27.03.2016)

[15] http://saturn.jpl.nasa.gov/multimedia/products/pdfs/20060328_CHARM_Webster.pdf
(retrieved 16.02.2015)

[16] http://www.jpl.nasa.gov/spaceimages/images/largesize/PIA03883_hires.jpg
(retrieved 23.08.2015)

[17] http://www.nasa.gov/images/content/446045main_landing_full.jpg
(retrieved 09.03.2016)

[18] http://spaceflight.nasa.gov/gallery/images/shuttle/sts-132/hires/s132e012208.jpg
(retrieved 16.09.2015)

[19] http://www.nasa.gov/sites/default/files/thumbnails/image/iss044e000028.jpg (retrieved 17.11.2015)

[20] http://www.astronautix.com/graphics/t/tt990515.gif (retrieved 24.07.2015)

[21] http://www.esa.int/spaceinimages/Images/2015/11/Trace_Gas_Orbiter_Schiaparelli_and_the_ExoMars_rover_at_Mars (retrieved 11.04.2016)

Further Reading

The textbook presented here has been written for aerospace students, scientifically interested individuals and employees in the aerospace industry. Considered as a compendium, it is intended to set out the foundations of aerospace technology. For reasons of better legibility, refererences to further reading were not always given throughout the text. A concise selection for further reading on each topic is offered in the following.

Appel F, Paul JDH. Oehring M. Gamma Titanium Aluminide Alloys: Science and Technology. Wiley-VCH Verlag, Weinheim, 2011.

Baker D. NASA Space Shuttle Manual: An Insight into the Design, Construction and Operation of the NASA Space Shuttle. Haynes Publishing Group, Sparkford, Somerset, 2011.

Dubbel H, Davies BJ, Beitz W, Küttner KH. Handbook of Mechanical Engineering. Springer Verlag, Berlin, 1994.

Elsner E. Raumfahrt in Stichworten. Ferdinand Hirt Verlag. Kiel, 1989.

Hale W, Lane H, Chapline G. Wings In Orbit: Scientific and Engineering Legacies of the Space Shuttle. CreateSpace Independent Publishing Platform, Charleston, 2012.

Hall R (Ed). The History of Mir. 1986 – 2000. British Interplanetary Society, London, 2000.

Hall R (Ed). Mir. The Final Year. British Interplanetary Society, London, 2001.

Hamann CH, Hamnett A, Vielstich W. Electrochemistry. Wiley-VCH Verlag, Weinheim, 2007.

Hohmann W. Die Erreichbarkeit der Himmelskörper. Oldenbourg Verlag, München, 1925.

Kainer KU, Magnesium – Alloys and Technologies. Wiley-VCH Verlag, Weinheim, 2003.

Ley W, Wittman K, Hallmann W. Handbook of Space Technology. John Wiley & Sons Ltd., Chichester, West Sussex, 2009.

Leyens C, Peters M. Titanium and titanium alloys: Fundamentals and Applications. Wiley-VCH Verlag, Weinheim, 2003.

Mark K, Kennedy GP, Joels DL. The Space Shuttle Operator's Manual. Ballantine Books, Westminster, Maryland, 1982.

Markley FL, Crassidis JL. Fundamentals of Spacecraft Attitude Determination and Control. Springer Science+Business Media, New York, 2014.

Messerschmid E, Fasoulas S. Raumfahrtsysteme. Eine Einführung mit Übungen und Lösungen. Springer-Verlag, Berlin, 2011.

Metzler R. Der große Augenblick in der Weltraumfahrt. Loewe Verlag, Bayreuth, 1980.

National Space Administration. NASA Facts. Landing the Space Shuttle orbiter at KSC. KSC release no. FS-2000-05-30-KSC, revised May 2000.

Oberth H. Wege zur Raumschiffahrt. Verlag von R. Oldenbourg, München and Berlin, 1929.

Prasad NE, Gokhale AA, Wanhill RJH. Aluminum-Lithium Alloys: Processing, Properties, and Applications. Butterworth – Heinemann, Oxford, 2014.

Rana S, Fangueiro, R. Advanced Composite Materials for Aerospace Engineering. Elsevier Science & Technology, Amsterdam, 2016.

Riley C, Dolling P. NASA Apollo 11: An Insight into the Hardware from the First Manned Mission to Land on the Moon. Haynes Publishing Group, Sparkford, Somerset, 2010.

Ruppe HO. Introduction to Astronautics. Volume 1 & Volume 2. Academic Press, London, 1966/1967.

Sagirow P. Satellitendynamik. B. I. Hochschultaschenbücher, Mannheim, 1970.

Sagirow P. Stochastic Methods in the Dynamics of Satellites. Springer-Verlag, Wien, 1970.

Schäfer H. Astronomische Probleme und ihre physikalischen Grundlagen. Vieweg Verlag, Braunschweig, 1980.

Siahpush A, Gleave J. A brief survey of attitude control systems for small satellites using momentum concepts. In: Proceedings of the 2nd AIAA/USU Conference on Small Satellites. Logan, 1988.

Schmidt R. Physik der Rakete. Mittler Verlag, Frankfurt am Main, 1963.

Sutton GP, Biblarz O. Rocket Propulsion Elements. John Wiley & Sons, Hoboken, New Jersey, 2010.

Index

www.ingramcontent.com/pod-product-compliance
Lightning Source LLC
Chambersburg PA
CBHW061415210326
41598CB00035B/6217